ISAMBARD'S KINGDOM

TRAVELS IN BRUNEL'S ENGLAND

JUDY JONES

SUTTON PUBLISHING

First published in 2006 by
Sutton Publishing Limited · Phoenix Mill
Thrupp · Stroud · Gloucestershire · GL5 2BU

British Library Cataloguing in Publication Data
A catalogue record for this book is available from the British Library.

ISBN 0-7509-4282-7

For Alison and Howard

Typeset in 10.5/14.5pt New Baskerville.
Typesetting and origination by
Sutton Publishing Limited.
Printed and bound in England by
J.H. Haynes & Co. Ltd, Sparkford.

Contents

Acknowledgements

Grateful thanks are extended to the many people who helped me to research and write this book. Firstly, to friends and familiars who accompanied or accommodated the dog and me on various sections of our travels: Anne Boston, Kathryn and Paul Robinson, Kathy Harvey and Dan Levy, Malcolm Rigg and Lesley Saunders, Mark and Tanya Hughes, Alison Griffiths and Howard Sprange, Sue Alexander, Angie and Huw Weatherhead, Roger and Elizabeth Meldon-Smith, Sylvia and Peter Christie.

Ian Christie's help in working up the germ of an idea into a potential proposition was much appreciated; as was my contact with Andrew and Melanie Kelly and their colleagues at the Brunel 200 project in Bristol in the spring of 2005. It was great to meet along the way people who gave generously of their time and interest, shared their local knowledge and often pointed me in the right direction. My mother, Jean Jones, funded the purchase of the digital camera I used to help record the journey.

I am also grateful to Simon Fletcher and colleagues at Sutton Publishing for their enthusiastic support and encouragement. May I also thank all the unsung heroes and heroines who keep all the preserved railways and railway and canal paths going, mostly in their spare time; and everyone who works to maintain and promote access to public rights of way, permissive paths, cycleways, bridleways, local and national trails; in particular the Grand Union Canal, Jubilee River, Thames Path, Ridgeway, Swindon Old Town Railway Path, Wilts and Berks Canal, Avon Walkway, Kennet and Avon Canal, Bridgwater and Taunton Canal, Grand Western Canal, Templer Way, Camel Trail, Saint's Way, South-West Coastal Path, Mineral Tramway routes and St Michael's Way.

Finally, I must acknowledge the contribution of the redundant farm worker who helped to keep me on track and in order throughout my travels – my dear border collie Fly.

Preface

Two centuries after the birth of Isambard Kingdom Brunel, the fruits of his inspired thinking and creative collaborations, the sheer force of this maverick personality remain stitched into the fabric of everyday life. His magnificent bridges and viaducts, tunnels and station buildings are part of our living, working landscape. Dozens, maybe hundreds, of public places across the south of England and the West Country – streets, shopping centres, pubs, car parks, a university, bus stations, tower blocks, even an entire forest – bear the name of Brunel or the Great Western Railway in recognition of his many astonishing achievements.

Isambard Kingdom Brunel (1806–59) was a dynamo of a man whose prodigious energy and prolific output of ideas, designs and completed projects were helpfully underpinned by an unrivalled self-belief. Canny and articulate, he could charm his adversaries as effectively as he could exasperate and undermine loyal colleagues, who sometimes knew better than he but lost the argument. He knew he was different, and he made sure others did too. 'The finest work in England' was his verdict on the GWR, years before a single length of track had been laid.

There was something of the savant about Brunel. He could barely sit still for five minutes without sketching the outline of some astonishing viaduct, bridge, station façade or flourish of castellated stonework. If he wasn't galloping his horse across the Thames Valley and the Wiltshire Downs surveying routes for his railway, or rattling around in his four-wheeled carriage-cum-mobile office visiting construction sites, he was despatching withering rebukes to recalcitrant contractors, awkward landowners and the like; or, when he allowed himself a rare moment's peace, designing the park and gardens of the retirement home he planned to build near Torquay – a dream he was never to fulfil. 'You and many others may think my life a pleasant one because I am of a happy disposition,' Brunel once wrote to his brother-in-law. 'But from morning to night, from one end of the year to another, it is the life of a slave.' In those few words, he made a shrewd and prescient self-assessment: Brunel died from exhaustion at the age of fifty-three.

The idea for this book grew out of a visit to Swindon's GWR railway village and Steam Museum in the summer of 2000. I was exploring the town on foot to research an article for the *New Statesman* about Swindon's aspiration to win city status, one which I felt at the time, as a result of my urban rambling, was doomed to remain unfulfilled. As I became more absorbed by the history of the railway, a question began to take shape in my mind and just wouldn't go away. Might it be possible to 'walk' the Great Western as well as ride on it, I wondered? On autumn and winter weekends, I started to devise and test out a walking route from Paddington to Penzance shadowing the GWR mainline, and some of its progeny of branch lines and extensions, along public footpaths, bridleways and byways. Having lived near the line for most of my adult life, I wanted to discover more about Brunel's legacy, and explore on foot the landmarks, landscapes and communities shaped by Brunel's extraordinary vision. As the granddaughter of a GWR signalman, whom I knew very little, I was also curious to find out more about his world.

Why walk when I could drive or take the train? Well, sitting on a train, especially one travelling at 100mph-plus, your view of railway architecture and infrastructure – bridges, viaducts, tunnel entrances and exits, cuttings, stations – and the hinterland of people and places beyond is decidedly limited or impossible. You don't really see the railway in the round either from a train or a car: the communities and workplaces thrown up to service it or to exploit the social and economic opportunities the line generated; evidence of the seismic revolutions sparked across social and class divides – gentry, aristocracy and factory workers – by the coming of the railway, blurring the boundaries of time and space.

What clinched the decision to attempt the journey on foot was the recent arrival in my life of a lively young border collie called Fly, a creature of almost boundless energy. I reckoned that a 500-odd-mile walk across six counties might suit her constitution as well as satisfy my curiosity. If I could persuade some of my friends to join us for various sections of the walk, so much the better.

The only other person I was aware of who'd walked from Paddington to Penzance was Charles George Harper, author of *From Paddington to Penzance: The Record of a Summer's Tramp from London to the Land's End* (Chatto & Windus, 1893), a whimsical and waspish account of his journey via Petersham, Shepperton, Windsor, Basingstoke, Winchester, Bournemouth, Weymouth, Exeter, Totnes, Dartmouth and Polperro. Accompanied on the trip by a friend identified only as 'the Wreck', Harper offers few clues in the book as to his motivation, and makes almost no mention of the railway. 'I do it for the reason why poets write poetry – because I must,' he wrote.

In a television poll for the the BBC's *Greatest Britons* series in 2002, Brunel gathered 398,500 votes, behind Sir Winston Churchill, elbowing Diana, Princess of Wales into third place. By this time, several landmarks along the GWR had been nominated for World Heritage Site designation.

Four years after starting to piece together a viable walking route, and four days before the summer solstice of 2005, I finally set off from Paddington station, with Fly on her lead, a bundle of well-thumbed OS maps and a backpack full of dog food. At last we were on our way.

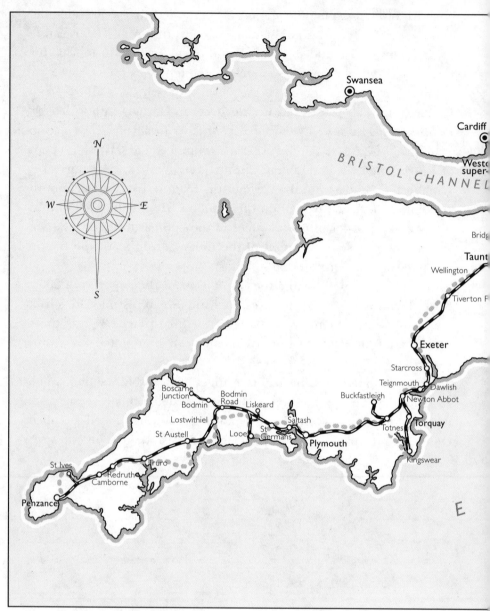

The route from Paddington to Penzance.

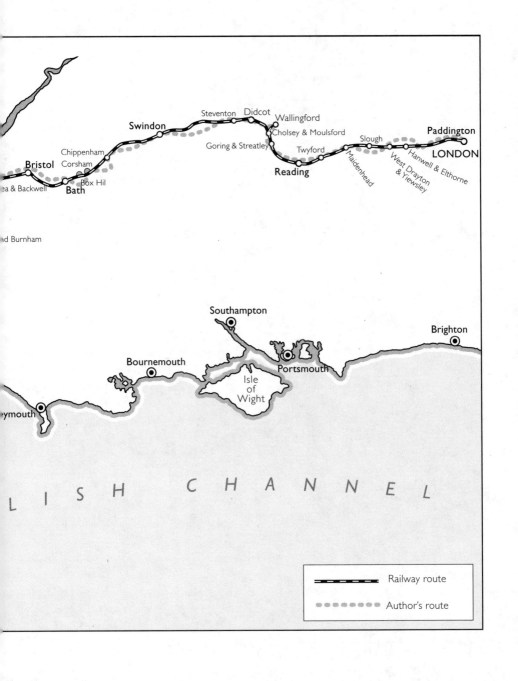

Steventon   Didcot
Swindon   Wallingford
Chippenham   Cholsey & Moulsford
Corsham   Goring & Streatley   Twyford   Slough
Bristol   Paddington
      Box Hil   Reading   LONDON
Bath   Maidenhead   Hanwell & Elthorne
ea & Backwell   West Drayton
                                                           & Yiewsley

d Burnham

Southampton

                                                    Brighton
Bournemouth

            Isle   Portsmouth
            of
            Wight

ymouth

L I S H   C H A N N E L

| ▬▭▬▭▬▭ | Railway route |
| •••••••••• | Author's route |

Three Bridges & a Statue

Paddington to Maidenhead

In which conversations with anglers are attempted, sandwiches are eaten at the cemetery, an accident is narrowly averted on the Grand Union Canal and we run out of water east of Horlicks Castle.

In a narrow road sunk beneath Eastbourne Terrace, on the western edge of Paddington station, black cabs queue to drop off their fares and pick up the next. People hauling wheeled suitcases, wearing backpacks, carrying laptops and briefcases, stream in and out of the entrance lobby to platform 1. Some hurry past the statue of a seated figure on a four-sided plinth in the middle of the entrance lobby, glancing in his direction purely to avoid a collision. Others pause to round up dawdling children or light a cigarette.

Marooned by this perpetual tide of human traffic and luggage, the sculpted figure appears uncomfortable in his chair. The booted legs are crossed at the knee; the hat is grasped firmly in his left hand, as though at any moment it might be carried away by a gust of wind. The vacant facial expression and the passive posture offer few clues as to the identity of the man depicted, or why this representation of him should be plonked in the middle of the entrance to one of London's major railway stations. While sharing certain common features – tall hat, long coat, booted legs – the statue of Paddington Bear is much better located, tucked within the more spacious area behind the escalators.

The image projected, to me at least, is that of a genteel early Victorian of modest distinction. Under Secretary of State for Canals and Turnpike Roads in the Peel administration perhaps or some obscure parliamentary draughtsman, remembered – if only by his close circle – for his mastery of Corn Law repeal or the Enclosures. 'Brunel greets you,' smiles my friend Anne Boston, waving a hand at the statue, waiting to join us for the first day of our walk, as far as Hanwell. Now it used to be a brisk ten-minute stroll, up Eastbourne Terrace, right on to Bishop's Bridge, then left down to the canal path and the sudden tranquillity of Little Venice. Today, as luck would have it, Bishop's Bridge Road has a massive great hole in it, where the bridge used to be. It has disappeared into thin air. My carefully worked out route from Paddington station down to the Grand Union Canal path is already shot to pieces and in need of revision. 'All you have to do is go back into the station,' a Welsh workman tells us, 'walk over the footbridge, follow the signs to Sheldon Square and Harrow Road. Then you'll be on the canal.' We pass some information boards along the way explaining the hitherto mysterious disappearance of Bishop's Bridge – it's being replaced with a wider, stronger structure.

The vast area north of Paddington station, around the canal basin, is a gigantic dusty building site, partly contained behind high padlocked boards, partly spilling out on to roads and pavements lined with traffic cones. The redeveloped Paddington will supposedly do for the western side of inner city London what the docklands revamp around Canary Wharf did for the eastern end, turning a long, dreary, neglected and forgotten area into a hub of gleaming offices, smart flats, regeneration opportunities and attractive public open space.

Reaching the waterside just short of Little Venice, a confection of coloured boats and lush greenery envelops the walker. Here you can almost hear yourself think and breathe again, as the sound of roaring beeping traffic is swiftly displaced by that of lapping water and birdsong. Splash-landing Canada geese scatter moorhens and coots in all directions, squawking and flapping furiously in protest.

The smell of frying bacon and percolating coffee wafts over the water to torture passers-by on the tow-path; children and dogs running loose on the canalside are ordered back to their moorings. Further along, peeling paintwork is removed and recoated, slowly, lovingly by boat owners as though embarked on the developmental stages of serious works of art. From here you can take a boat ride to Regent's Park or Camden Lock, or just sit at the Waterside Café and watch the world amble by. Today the peace of it all, the deep blue sky patched with milky-white clouds and waving treetops above, is almost intoxicating.

'Time for a coffee, Anne?' I suggest. Having let the dog off the lead, and eyed the floating café, I am beginning to feel dangerously relaxed. Noting that we have only been walking for about fifteen minutes and still have about 12 miles to walk before finding the Wharncliffe Viaduct, our day's destination, Anne replies: ' Perhaps we ought to press on a bit?'

The Grand Union – Britain's longest canal – opened in 1800, providing a 137-mile waterway linking the rivers Trent and Thames, from Birmingham to the London docks. The poet Robert Browning coined the phrase Little Venice to describe this particular area, around the junction of the Paddington arm of the Grand Union with the Regent's Canal. His home overlooked the canal, and curiously enough he died while on holiday in Venice (the Italian one) in 1889. I am loath to leave such a cheery place, where the boats have tomato plants and herbs growing in pots on tiny foredecks, and crocheted porthole covers. Amid billowing clumps of cow parsley, yellow flag iris and bramble, we continue along the path, as the white stucco Regency villas backing on to the northern side of the canal give way to new apartment blocks.

We pass the austere-looking Trellick Tower, once one of the most loathed buildings in London. Built in the brutalist style of architecture popular in the 1960s, the block fell out of favour for many years, but is now highly desirable among young affluent professionals.

Having endured the thundering traffic above us on the Westway flyover, and passed the concrete walls topped with razor wire, it's a relief to reach Meanwhile Gardens, a wonderful linear park of nature walks through mature trees, bulrushes and its own little brook and skatepark.

The balconies of the flats on either side of the canal around here are stuffed with all manner of domestic paraphernalia – bikes, bird-cages, drying washing and buckets.

We soon turn off the canal path to visit Kensal Green Cemetery, the first necropolis to be built in London, opened in 1833, and still owned by the company that founded it. Similar cemeteries were being laid out in Liverpool and Glasgow around this time as church burial grounds became ever more crowded. Back in 2001, I'd joined one of the regular Sunday afternoon tours organised by the Friends of Kensal Green Cemetery.

About a dozen of us had gathered at the Anglican chapel and were led down into the catacombs where the remains of the dead are held in coffins beneath numbered arches. I remember our guide, Henry Vivian-Neal, pointing out that the choicest spots in the catacombs are located at head height for the convenience of visiting relatives and descendants. The shelves above these tended to be reserved for children: the relative lightness of their coffins enabled them to be pushed into position more easily than those accommodating adults.

The cemetery's published guide and list of fifty notable monuments reads like a Who's Who of eminent Victorians and late Georgians. William Makepeace Thackeray, Wilkie Collins, Anthony Trollope are here. So too is Blondin, the famous tightrope walker who once paused to cook and eat an omelette while balancing on a rope high above the Niagara Falls; the Revd Sydney Smith and Princess Sophia, daughter of George III. The remains of more than 200,000 individuals are buried here in 63,000 graves, spread over 77 acres, a grey-green jumble of chiselled stone, mature trees, wildflowers and neatly mown paths. It's still used for burials, with plenty of newly dug graves and fresh flowers on them, and people filling watering cans from taps, sprucing up the plots of departed loved ones. All around us are towering mausoleums, Greek temples, Gothic extravagances, topped by triumphant stone angels, arms outstretched.

The monument to the civil engineering father and son, Sir Marc Brunel and Isambard Kingdom Brunel, and their families, is easily missed. About 3ft high, the stone is just big enough for the inscriptions naming those commemorated: Sir Marc and Dame Sophia Brunel; their son Isambard Kingdom Brunel and Mary his wife; their children Isambard and Henry Marc; a daughter-in-law and a niece. The simple stone monument, unadorned and facing west, is contained within a plot of about 12ft by 8ft, which is edged with stone, and covered with grass and weeds. Shaded by an ash tree in full leaf, it's a rather lovely spot – a little north of the Grand Union Canal, a pair of huge gasholders and beyond it Brunel's own Great Western Railway. The air is filled with birdsong and the occasional rumble of a passing train. 'Shame about those weeds,' I say, as we sink to the grass for a few minutes break and a sandwich.

'I don't think Brunel would have minded a bit about the weeds,' Anne replies, flicking a crumb off her trousers. 'His memorials are all around, those marvellous bridges and viaducts. They're the ones that matter. He wouldn't have minded about the weeds.'

We rejoin the canal path next to Sainsbury's. A moorhen with her brood is nesting on a plank of wood, slowly circling the surface of the water. Anne excels at noticing and identifying the wildlife – common terns and cormorants, here, bee orchids, mallow and a water vole, there. It's a sweltering hot day and we have to keep ourselves, and the dog, topped up with water at regular intervals. The dry air suddenly fills with the pungent aroma of curry spices drifting our way from some open doors and windows on an industrial estate. From the aqueduct over the North Circular Road, we pause to glance down at eight lanes of roaring traffic and move on swiftly. Within seconds we're back among the lush green vegetation, dragonflies and waterbirds, and peace. The canal seems to provide a blissfully slow-motion parallel universe, a twilight zone of anglers, walkers, boaters and cyclists taking life at a relaxed pace.

Anne and I try very hard to avert our gaze from a bare-bottomed angler. Although he appears to be wearing shorts, they are so far down his body they might as well not have been there. We creep past nonchalantly, being women of the world, as though seeing such vast expanses of bare flesh was an everyday trifle. Like many anglers, he sits in complete silence staring into the mid-distance, occasionally fiddling with a margarine tub full of maggots. It's a man thing, we conclude.

We manage to converse with a fully clothed angler further on. 'Nice bike,' I venture, nodding at the two-wheeler propped against a mesh fence. 'Should be for 650 quid,' he replies, continuing to contemplate the mid-distance.

'Battery-powered?'

'Yup. Charge it up every 30 miles or so.'

'And plenty of room for all your tackle . . .'

'That's right. It's a good bike,' he says.

'So – you don't have to pedal up all the hills then?'

'That's it – you got the choice. Pretty good bike.'

Having satisfactorily ascertained most of the bike's salient features, we take our leave. Further down the path, the landscape assumes a more suburban open feel, a mix of semi-detached houses and low-level blocks of flats set in gardens with neatly trimmed lawns. The gaps between buildings seem to get bigger. We wave at some children having a birthday party on a canal boat festooned with red, white and blue balloons. At Perivale we cross the footbridge over the Westway and, leaving the traffic noise and exhaust fumes behind us, slip through a lych-gate towards the tiny weatherboarded St Mary's Church and into the Brent Valley River Park.

A family of Canada geese patrols the fairway as the late afternoon shadows lengthen across Brent Valley golf course. The brambles are in full flower, and the heads of the giant hogweed standing to attention on either side of the river path are drooping and turning yellowy-white. 'They've been sprayed with weedkiller,' Anne says authoritatively. Soon we come to a glorious yew maze enclosed by picket fencing. Children are chasing each other round it to see who will be first to the tree house in the middle.

Anne has been sporadically troubled by the complete absence of female mallards all along the route from Paddington – plenty of males but no females. 'Look, another group of gay mallards,' she ventures.

I'm more troubled by the continuing absence of the Wharncliffe Viaduct. 'It's here somewhere Anne, I assure you. Must be just behind this little bit of woodland and that flock of magpies.' We pick our way through the trees, heading south towards the Uxbridge Road and Ealing Hospital. The magpies fuss and flap across our path, as I wrestle with the OS map.

'It's a *parliament* of magpies,' Anne says. Suddenly we emerge from the dappled sunlight of a small thicket to find ourselves in darkness, the temperature a degree or two cooler, as if in the midst of an eclipse, a few feet away from the base of one of the gigantic brick arches of the viaduct towering above us. It's an awesome sight.

Nearly 300yds long and 65ft high, the viaduct was built to carry the railway over the Brent river valley. You have to tip your head right back to see the top. The Brent river path takes us right underneath one of the easternmost arches, perfectly illuminated by the shimmering reflection of the late afternoon sunshine on the water. Trees grow out of the arches, underneath the arches, and all around them. It's not until you cross the hay meadow to reach the Uxbridge Road and Ealing Hospital, to look back at the viaduct from a distance, that you appreciate its great length. Hundreds of men and horses worked all along here between 1836 and 1837 to construct the viaduct, Brunel's first major structural design, the first contract to be let for the GWR, and the largest brickwork construction along the 118-mile main line from Bristol to London.

We join the queue at the bus stop opposite the hospital, me heading for my overnight stay with friends in Ealing, Anne heading home to Kentish Town, north London. We shoehorn ourselves and the dog on to the rush-hour bus, and it takes quite a few minutes for us to persuade Fly to leave her spot under a seat to disembark at Ealing Broadway.

* * *

My friends Kathy Harvey and Kathryn Robinson and I all used to live in Ealing back in the 1980s and early '90s, but of the three of us, only Kathryn is still there. We're sitting on top of a 607 bus from Ealing Broadway, back to the Hanwell viaduct for the start of the second day's walk to Slough.

We pass the late lamented Daniel's department store – for decades a Mecca for pregnant women across west London and a landmark along the Uxbridge Road. The familiar hoarding attached to the front of the building promising '400 PRAMS IN STOCK' is still in place, but unfortunately the store is empty – awaiting redevelopment. Both Kathy and Kathryn, who shopped for their children there, mourn its passing. Back on the river path, this time south of the Uxbridge Road and Ealing hospital, the rural areas of Hanwell – which I'd long assumed had disappeared under concrete – are surprisingly much in evidence. There are pretty cottages, vast allotments, pockets of rural London which I never knew still existed. Within minutes we're back on the Grand Union Canal path and passing a series of beautifully maintained locks.

Despite the continuing heatwave, the dog is in her element, slurping water from her collapsible canvas travelling bowl and lying down to conserve her energy whenever we all stop to examine some detail of canal infrastructure. 'That's a true collie,' says a man with two of his own at his feet, nodding in Fly's direction. He spots her weaving back and forth behind me. 'She's rounding you up!'

We pass the tall back wall of St Bernard's Hospital, originally the Hanwell Lunatic Asylum built in 1831. In this wall bordering the tow-path small doors were built so that fire hoses could be put into the canal in an emergency. Within a mile or two of passing Brunel's first major project design for the GWR, we soon find ourselves bang in the middle of one of the last he engineered. Directly above our heads is a road bridge, and we're standing on the aqueduct taking the canal and tow-path over the railway line below us. It's a very odd feeling, as if we've been pitched into one of Escher's paintings of impossible architecture. Kathryn, a planning officer who knows a thing or two about architecture, is examining part of the stone structure supporting this intersection between road, canal and railway. 'These are revetments,' she says.

'They've just listed similar ones at South Kensington tube station.' What we're looking at, indeed are surrounded by, is Windmill Lane Bridge, or Three Bridges as locals call it. It was as a result of the need to avoid routing the planned Great Western and Brentford Railway through nearby Osterley Park that work on the unusual construction was started in 1856.

Hanwell's flight of six locks raised the level of the canal 53ft in little over a mile and, like Windmill Lane Bridge, is a scheduled ancient monument. A school sports day is getting into swing on the other side of the canal. 'Look out for the carp jumping after the next lock,' says a walker we meet going in the opposite direction to us. A family of coots sits on a bit of old fencing panel bobbing up and down on the water. Anne had told me that coots make the cruellest of parents – they feed only one of their young, the one they judge to be the healthiest. The rest have to fend for themselves, apparently. I pass on this intelligence to the others, but Kathy is more interested in the dazzling golden orb she spies through some trees in the direction of Southall.

At Bull's Bridge, we turn right and head north, soon passing some double-decker houseboats and a Tesco Extra on the other side of the canal. We smell the Nestlé factory before seeing it, the whiff of instant coffee assailing the nostrils, unleashing a bout of sustained reminiscing about our respective upbringings in the late fifties and sixties, when our mothers first experienced the emancipating effects of convenience foods and stuffed our larders with them. 'When I were a lass, a cup of Nescafé instant coffee with a spoonful of Marvel dried milk was the height of sophistication, especially when accompanied by a fig roll,' I say.

'Aye, luxury,' says Kathy. 'Especially after a plate of rehydrated Vesta Chow Mein, followed by a bowl of Angel Delight.' Halcyon days. In the semi-darkness underneath the railway line, the canal path narrows to single-file width. Noting that the dog is about to fall into the water, I dive forward to grab her by the scruff of the neck just in time.

Eventually we see some carp jumping, as promised, either oblivious to the fast-food packaging and plastic bottles littering the canal on the stretch up past the Lake Farm country park and the Glaxo factory, or perhaps in protest at it. After passing West Drayton station, and some smart canalside apartments, we miss the first turning off to the Slough arm of the Grand Union Canal. A dog walker directs us to the right footbridge. 'Your dog's got a bit of collie in him, hasn't he?' the man says.

'More than a bit, ' I reply. 'It's hard to know – she came from a farm in Herefordshire. Not quite sure about her breeding.' Having said our goodbyes to Kathryn at Southall, Kathy and I and the dog slump under the shade of some trees opposite a boatyard. We find that we've run out of water, so I hop over the road bridge carrying two empty water bottles, heading for the boatyard. 'Have you a drinking water tap, so I can fill up our bottles?' I ask a chap walking across the boatyard. 'We're walking on the canal path. The dog's a bit thirsty.' Without speaking, the man takes the bottles, disappears for a few minutes, and then reappears with replenished water bottles. 'Thanks so much,' I say. 'You're very kind.' He hands them back, again in silence, and returns to his duties.

Next to some of the narrow boats along the Slough Arm, there are rotary clothes dryers, plastic chairs and tables, but very few other signs of life. Dotted about on the canal's surface, some water lilies are coming into bloom, with small white flowers opening or about to open. The water bottles, full only an hour or so ago, are now empty again. In the distance straight ahead, I spot my friend Malcolm Rigg pedalling along the towpath to meet us. We first met about twenty years ago and, as always, it's great to see him again. Having introduced him to Kathy, and reminded them that she lives in the same village as his sister – Eynsham, near Oxford – I get down to basics. 'Malcolm, have you got any water with you?'

There's not much he doesn't know about Slough that's worth knowing, and as we continue to head into town, Malcolm starts telling us about its industrial history: 'When the GWR was planning the railway, the decision was made that there wouldn't be a station at Slough, so instead they built it at Langley. This was because there had been objections from Eton College – just up the road – that putting the station in the town would enable the boys to escape too easily.'

I later learned that the Provost of Eton himself had attended a public meeting at Salt Hill to voice his concern. 'If the boys could be carried to a distance of five miles in fifteen minutes, they could easily put themselves out of the reach of the authorities, and so the school must be injured,' he told the meeting. In June 1838, Eton College applied for a court injunction to prevent the GWR stopping at Slough. The application was dismissed.

Resuming his story, Malcolm continues: 'Slough was just seven parishes – with Upton in the middle. That's where the trains stopped.

In the end they did build a station here, which opened in 1840. They built a lovely hotel next to it, the Royal, but that didn't last long. It closed pretty rapidly afterwards. As soon as they built the branch line to Windsor, they no longer needed a railway hotel in Slough because people coming to stay in the area were often visiting royalty.' On the site of the Royal Hotel now stands a modern building housing the mobile phone company O_2. Next to it a vast new Tesco supermarket is nearing completion. Many of the houses in Herschel Street – where Malcolm and his partner Lesley Saunders live – were built for servants at Windsor Castle. Slough had been a coaching stop, having the Bath Road passing through it. There were lots of coach houses here. Curiously, the Slough Arm of the Grand Union Canal wasn't built until 1882, long after the railway had arrived.

Malcolm continues: 'Although the golden era of the canals died with the coming of the railways, much of the land around here was brickfields and so, unusually, the Arm was built because it was needed to transport bricks.' Pointing to the field next to us, he carries on: 'So you can see there, for example, the level of the land is quite a bit lower than it is here on the canal, and that's because the top 6ft or so of clay was being extracted to make the bricks.'

Passing by the ICI Dulux paint factory on one side, and a row of moored narrow boats on the other, we soon reach the Slough Basin. An information board here confirms that the Slough Arm of the Grand Union Canal was one of the last canals to be built in the country. Initially, bricks were carried by barge from Slough, Iver and Langley to London. When the brickfields were exhausted, sand and gravel became the main cargo. Rubbish was transported back out from the capital to fill the gravel pits beside the canal. By the 1950s, the canal trade had been lost to road transport and the canal became run down. Now restored, the waterway provides a recreational route and an important refuge for wildlife.

As we walk down the road towards Slough's railway station, Malcolm points out Horlicks Castle, crowning the factory where the famous malted bedtime drink is made, and the roundabout immortalised in the opening sequences of the TV series *The Office*, Ricky Gervais's sitcom set in Slough.

By suggesting its wholesale obliteration, John Betjeman, the late poet laureate, put twentieth-century Slough on the map, much to the chagrin of loyal locals. 'Come friendly bombs and fall on Slough,/It isn't fit for humans now . . .' The irony is that his famous poem was first published in 1937, long before the soulless, ugly redevelopment inflicted on certain parts of the town in the 1960s was even contemplated. Slough eventually retaliated with a collection of its own poems, 'In Praise of Slough', edited by the performance poet Attila the Stockbroker. One of Lesley's poems, 'Going to the Dogs', is featured in the collection, inspired by the sale of the greyhound racing stadium to the Co-op.

'Is Station Jim still on platform 1?' I ask a ticket controller at the barriers.

'Yes, still there,' he replies.

'I saw him this morning,' adds Kathy, before rushing to catch the train back to Oxford.

Slough station was the first along the GWR to have both an Up and a Down platform. Up east towards Paddington, Down west towards Bristol. Most of the original station is intact, particularly the Victorian cast ironwork and the weatherboarding, now decorated with cheerful hanging baskets. Station Jim was a hound of indeterminate breeding that used to pad about the place, collecting for the GWR widows' and orphans' fund. He wore a harness with leather pouches on either side for collecting donations, still in place on his stuffed remains, enclosed by a glass case.

* * *

Sunday morning. The heatwave goes on, and the insect bites on my leg have grown into marble-sized blisters. Lesley suggests I nip round to the NHS walk-in centre at Upton Hospital, a five-minute walk away, so off I go to join the queue of people waiting for the clinic to start. After the blue doors open at 9 a.m. about a dozen of us walking wounded hobble into the reception area. Each one is handed a form to fill in, and one of the questions asks why we are in the area. Am I a resident, a tourist or visiting for work purposes? Although I look and feel like a tourist, sporting the 'shorts and sunburn' look, I tick the box marked 'Work'. After just a ten-minute wait, I'm called into a treatment room, requesting aspiration. The nurse duly pops a sterile needle through my blisters, dresses the wounds and tells me to come back if I have any more problems. What an excellent service.

It's far too hot to walk to Maidenhead today, and Fly could do with a decent rest in the cool of Malcolm's house. Malcolm disappears into his garage and emerges with his tandem.

'I'm not sure about this,' I say, having never ridden tandem.

'There's nothing to it. You'll be absolutely fine.'

I must report that the back-seat rider's first ten minutes or so on a tandem are not happy ones. Mine weren't anyway. Your only task is to pedal, but otherwise relax and enjoy, but the instinct to try to steer and control the bike is extremely hard to resist. As a result, my fidgeting about on the saddle is having a dangerously destabilising effect on the bike.

'Don't do anything,' he says. 'Just pedal when I do and relax.'

'That's all very well, Malcolm, but I think I'd like to go back now.'

Soon we're heading south out of Slough town centre through Herschel Park towards the Jubilee river, which we will follow to Maidenhead. By now I'm beginning to lose my beginner's wobble and we're speeding along pretty smoothly. It helps to shut your eyes, try not to think about anything, apparently. Pedal and relax, pedal and relax, I chant to myself. Eventually my fear of crashing to the ground abates.

The Jubilee river is a kind of Thames bypass. It was dug out of farmland and waste ground between the M4 and Windsor to alleviate serious flooding problems in and around Slough, Maidenhead and Eton. Meanders were built into the design, so it looks like a natural river rather than a man-made one. Thousands of trees have been planted on its banks, and sandbanks created in the middle of the river to create habitats for birds. It leaves the Thames at Maidenhead and rejoins it about 7½ miles downriver at Wraysbury. 'Its capacity is greater than the Thames – it will take a greater flow of water,' says Malcolm. 'It's fantastic to have this new river because the whole area is now a wetlands, and it's brought us a different way of getting about without having to travel by road. Another benefit of the Jubilee river path is that it gives walkers and cyclists an alternative to the Thames path, which gets pretty muddy in the autumn and winter and congested in the spring and summer.'

Improvements to Slough's sewage treatment works have been incorporated into the scheme, which was carried out in part to help mark the Queen's Golden Jubilee. In Slough, the new waterway is known also as Herschel Waters – after Slough's most famous resident, Sir William Herschel, Astronomer Royal to George III. He used to live at the junction of the road where Malcolm and Lesley live, Herschel Street, and Windsor Road.

'That's where he built the biggest telescope in the world – 40ft. The house he lived in with his sister Caroline, a fairly simple affair, remained until the 1960s, when the bloody council gave permission to have it demolished. It's now been replaced by offices. There's a memorial to him there, and the Herschel Arms pub sign features a picture of William Herschel with a telescope.'

We pedal underneath the single-track Slough to Windsor branch line, past a long row of newly planted willow. Soon we can see Eton Wick and the magnificent outline of Windsor Castle over to the left, and the rather less inspiring Asda at Cippenham to the right. At Manor Farm weir, masses of cormorants are perched in a neat line along the barrier across the river as if waiting for someone to fire a starting gun. Looking north beyond the weir, we see the chimneys of the power station rise above the Slough trading estate. At the Dorney wetlands, there are plenty of bird hides and wooden platforms, and quite a few walkers milling about. The habitats created include scrub, reed beds, grazing marsh and deep water, designed to attract a diverse range of birds and other wildlife all year round.

From the A4 at Taplow, Malcolm takes us to see the new redbrick bridge that's been built to carry the railway over the new Jubilee river.

'That bridge was created without a single train being stopped,' says Malcolm. 'What they did was to freeze the soil rock solid and drill it out. It's quite astonishing really. This was the technology used by the Soviets to build the Moscow subway system.'

The whole area is a bridge-watcher's heaven. Soon we arrive at the Maidenhead Bridge and look at it firstly from the older road bridge. Lacking classical proportions, what is arguably Brunel's famous bridge may not seem particularly elegant. The interest lies in the engineering – and Brunel's almost obsessional quest to create here the flattest possible arches. His design was built in 1838, consisting of two brick arches of 128ft each, with a rise of only 24ft. Some predicted that the bridge would collapse the moment the scaffolding was removed, but their scepticism proved unfounded. The bridge was enlarged between 1890 and 1892 to accommodate more railway track.

Having biked along the river path to test out the Sounding Arch, Malcolm and I then head up the Thames path just north of Boulter's Lock to see the junction of the river with the Jubilee river, and then set off towards Windsor via Dorney Common. We stop to look at the very simple and attractive St Mary Magdalene Church, Boveney, owned by the Friends of Friendless Churches since 1983, now undergoing major restoration with the help of English Heritage, and much hard fund-raising work ahead to complete the repairs to the tower.

Windsor is thronging with people lazing about by the river, dangling their toes in the cool water, eating ice creams, basking in the late Sunday afternoon sunshine. We return to Slough via Eton College playing fields. I fell off the tandem only the once.

2

Cutting a Dash along the Thames
Maidenhead to Didcot

In which we discover the best place to view the Sonning Cutting, meet a whippet lover, shelter underneath bridges at Moulsford and Goring despite opposition from the dog, experience Cholsey's Wonderful Railway – the CWR – and the Didcot Railway Centre.

'Sorry Madam – dogs aren't allowed in here,' says a security guard patrolling Slough's Queensmere Shopping Centre, shooing us back to the entrance, with outstretched hands. Fly and I find an alleyway skirting the side of the shopping mall and soon we're passing the Brunel Car Park to reach Brunel Way and Slough railway station.

Having already cycled to Maidenhead, I thought I might catch a train and walk from there to Reading. Since it's another scorching hot morning, I decide to get the train to Twyford – one stop on from Maidenhead – and start the day's walk from there.

At Twyford the first job is to find a map seller. My OS Explorer map of the area has mysteriously disappeared, and if I'm going to get us safely to Reading by the end of the afternoon, I need to buy a replacement. Then there's the small but important task of identifying and locating the best bridge from which to view the Sonning Cutting. From memory, there are at least four possible options. Having spotted H.F. Newberry, cardsellers and stationers, in London Road, and parked the dog outside the shop, I venture in.

'I'm after an OS Explorer 159,' I explain to the two men serving in the shop. 'Reading, Wokingham and Pangbourne – the one for this area and Sonning.' One of them goes to the front window and squats down to rummage through their stack of maps and guides.

'Sorry – we're out of stock for the Explorer 159,' replies the assistant. 'But we've got the Landranger 175 – Reading and Windsor,

Henley-on-Thames and Bracknell, if that's any use to you.' He stands up, brings it over to me, spreads it out over a table full of books and gives me a quizzical look: 'What are you looking for exactly?'

'Well, I need to find the best bridge for viewing the Sonning Cutting, and then work out a way of walking on to Reading,' I reply casually. Recognising me as a connoisseur of railway cuttings, with a thoroughly worthwhile day's objective, the shop assistants leap into action. While one busies himself covering an entire sheet of A4 with detailed handwritten directions, the other grabs a pencil and starts to sketch the route I need to take on a Landranger 175 that I haven't even bought yet.

'This is the bridge you want, just here,' he says, pointing at the spot with his pencil. 'That's the highest of the bridges over the cutting and it offers the best view in each direction.'

Armed with the marked Landranger 175 and the sheet of instructions, I pay up, thank the assistants for their help and return to the dog. An hour or so later, having woven our way through Charvil, turned left along Park Lane, gone under the railway line and over Duffield Bridge, we pass another Brunel Way, this time on a modern housing estate. We skirt Sonning Golf Club, and the large detached houses in West Drive. Suddenly I emerge on to a busy A road, part of a noisy roundabout, where three Network Rail men are staring down at the Sonning Cutting and conversing in muted tones. I too stare down at the cutting – evidently from the wrong bridge. Somehow I've managed to overshoot. Ahead of me, looking east, and slightly above eye level is the right one – spanning the railway about 200yds down the tracks.

Retracing our steps along West Drive, we pause for drinks of water and a bit of shade from the midday sun. It's a private road of large detached houses. The front gardens of immaculate trimmed lawns are so vast you'd almost need to thumb a lift to the front doors. The garages alone (some properties have three) could accommodate an average-sized family. Eventually, we find the narrow lane – Warren Road – that leads to the right bridge, away from the noise of the main road. Standing at the middle of the bridge, looking down at the cutting, the only sounds come from passing trains and birdsong. My shop assistants at Twyford clearly knew their stuff, as you can well appreciate the depth and breadth of the Sonning Cutting from here.

Some 2 miles long and 60ft deep, the cutting took more than 1,200 men three years to excavate. Although both rural and urban industries were increasingly undergoing mechanisation in the early to mid-nineteenth century, it's worth remembering that the creation of the railways was overwhelmingly laborious manual work, carried out by navvies (short for navigators) using picks, shovels and gunpowder. The painstaking and repetitious nature of this work, however, had its occasional compensations for the navvies themselves, but more significantly for the emerging science of geology.

Deep excavations such as that required to create the Sonning Cutting enabled geologists to begin systematically gathering samples and recording information about the mineral wealth of the country. *The Quarterly Journal of the Geological Society*, first published in 1845, regularly reported findings at railway cuttings, quarries and tunnels, although navvies would sometimes pocket the fossils and take them home to their families before the geologists could get their hands on them.

In 1859, J. Phillips, president of the British Geological Society, said the value of the railway construction was 'strongly felt by every geological reasoner who touches on problems of the distribution of oceanic sediments, the boundaries of land and sea, the mixture and alternation of fresh and salt water, or the local origin and geological diffusion of particular forms of life.'

Heading down towards Sonning village, we pass Bluecoats School. In Saxon times, Holme Park, where the school now stands, was the seat of the Bishop of Salisbury. In St Andrew's churchyard, beautiful mature trees – copper beech, yew and laurel – shade the gravestones. We stop for some lunch and relief from the fierce heat in the garden at the Great House. While ordering from the bar, I asked for some water for the dog. This arrives in a white china bowl, served on a white china plate with a neatly folded pink serviette. They're very civilised in Sonning.

Refreshed and revitalised, we stride on to the Thames path and head west past Sonning Lock and a line of stationary motor cruisers, engines purring, waiting to pass through the lock gates. Across the meadow, the large modern buildings of the Reading Business Park, dominated by the Oracle empire, rise above the tall grasses.

'We came to Reading prepared for anything but charm in that town of biscuits, and we were not inclined to alter our ready made opinion upon sight of it,' wrote Charles George Harper in 1893, clearly not a great believer in the notion of travel broadening the mind.

The water's edge is shallow enough for the dog – a reluctant swimmer – to splash about without getting out of her depth. Some scaffolding has been erected for the Henley Regatta – due to start in a couple of weeks time. A cormorant perched on top of a dead tree monitors the riverside activity.

There are plenty of sunbathers and joggers about. Some sturdy benches, featuring quotes from Kenneth Grahame's classic *The Wind in the Willows*, soon appear at regular intervals along the riverbank. I fall into conversation with an elderly man sitting on a bench. He wears white shorts and boating shoes, and his bike is propped against the back of the bench.

'Going far?' he asks, warily eyeing my backpack and dishevelled appearance.

'Penzance,' I reply.

'Not sure you'll get there today,' he laughs. I explain the purpose of my mission, and he leans back against the back of the bench, folds his arms across his ample stomach and smiles ruefully. 'Brunel – now he was a real can-do genius,' he sighs. 'God's Wonderful Railway they called it, and it still was when I was a boy growing up during the war. The trains kept going during the bombing, some of them did. And if there was a delay, even of a couple of minutes, the guard would pass through each carriage and apologise to the passengers.'

We introduce ourselves. He's called Roy, and lives over the river in Caversham. Correctly anticipating a conversation of considerable length, Fly settles herself between Roy's feet and dozes off.

'We invented the railways, yet now we have an appalling railway system. You should see the ones in South America – much better than ours. Labour came in with this vision of an integrated transport system – all that seems to have gone out the window,' says Roy, warming to his theme. 'The trouble is, it's not run by railwaymen any more, but by accountants and politicians. Bob Reid, he was a railwayman, and the last one to run the railways properly. It really upsets me, just thinking about it . . .'

'Indeed,' I mutter. 'Well, we'd better get on . . .'

'Now I've got a senior citizen's rail card – you get a third off the ticket price if you travel after 9.30 a.m. or thereabouts . . .' Roy continues. A family of moorhens sails past us and I shield my eyes from the sun. 'The other day I was at Reading station at 10.15, about to board a train for London. A guard sees my travel card and says: "That's a travel card is it?" Yes. "You can't come on this train on that travel card". Why not? "Because this is the delayed 9.25 train, and you can't use the travel card before 9.30 a.m." Yes, but it's 10.15 now . . . "Sorry – you can't come on this train on that travel card." Honestly – bureaucracy gone mad.'

Catching sight of a common tern flying overhead, I sympathise with Roy's reflection on the incident. I stand up, waken the dog and heave the backpack over my shoulders: 'Well, we'd best be off.'

'Check out some of the journey times now,' he adds. 'Lots of the trains are slower now than in Brunel's time, I think you'll find.' We wish each other a pleasant afternoon and begin to take our leave. 'Look out for the mouth of the Kennet, next exit from the path into Reading,' Roy adds.

* * *

At the Thameside Promenade near Caversham Bridge, a coachload of senior citizens is climbing aboard the double-decked *Caversham Princess* for a gentle cruise along the Thames. Fly and I strike out on the path.

On the Caversham side are some beautiful houses, secluded by woodland, with their own moorings. There's one with a thatched and weatherboarded boathouse, with pale blue doors opening on to a balcony, and further upstream another sporting its own helipad (and red and white helicopter). Further along, just off the footpath on our side of the river, is a sumptuous Egyptian-style extravagance, painted in at least six different colours.

Sitting at Mapledurham Lock, watching the two lock-keepers working the locks under silver Environment Agency umbrellas, tending the gardens during gaps in the boat traffic, or selling plants to boaters, it's easy to lose track of time. One lock-keeper kindly fills up my water bottles and asks where we're headed. 'You should enjoy the next stretch, walking up to Pangbourne,' he tells me. 'Personally, I reckon the bit after that – to Goring – is even more attractive, because the path takes you up along the hillside and you get good views of the river through the trees. Plenty of shade too, if it's another hot day.'

There's something about being near rivers that encourages indolence and reverie, tendencies that are only exacerbated by the sun beating down on us. Walking can so easily turn into ambling, plodding even, especially towards the end of the afternoon. The area is rich in literary associations. Nearby Hardwick Hall is said by some to have inspired Toad Hall in *The Wind in the Willows* (although others say it was Mapledurham House); and the woodland above the north banks of the Thames, between here and Goring, the Wild Wood. This is all supposition of course. Kenneth Grahame spent his retirement at Church Cottage, Pangbourne, one of whose pubs – the Swan – is mentioned in Jerome K. Jerome's *Three Men in a Boat*. Both authors were born in 1859, the year of Brunel's death.

This is the sort of useless information the mind turns over at the end of a hard day's walking in hot weather. Having reached Pangbourne, it appears the dog's lead has gone missing. Back we go a mile or so to retrieve it from the grass, and head for the home of old family friends Sylvia and Peter Christie who are kindly putting us up for the night.

* * *

The weather remains warm and muggy as we set off from Pangbourne to cross the river into Whitchurch-on-Thames where we rejoin the path. The softly undulating wooded landscape between Pangbourne and Goring makes for a pleasant stroll. We walk high above the river for an hour or so through trees, along a path cut through the limestone and flint, slowly descending to the level of the water. The sounds of birdsong and distant trains, and wash from the occasional motor cruiser lapping the riverbanks, are all we hear for much of the day until the storm breaks.

As the skies darken, we pass under Brunel's Basildon Bridge and find shelter from the pouring rain beneath the road bridge at Goring. Watching the Goring lock-keepers work the gates, taking up their positions under red and white umbrellas, and the mists descend over Streatley Hill passes the time. The dog monitors every movement before draining almost an entire puddle. When a lock-keeper furls his umbrella, I realise it's time to move on.

At the Beetle and Wedge at Moulsford, packed out with suited diners, we have to come off the path and walk through the village along the pavement. Near the church, I hear the Cranford House School orchestra strike up 'I Vow to Thee My Country', playing not at all badly, while on the other side of the road, the Moulsford Preparatory School boys are playing rounders on their playing fields and volleying questions to 'Sir'.

Dropping back to the river I have a quick chat with an architect walking along the footpath who says that some of the land purchases made for railway construction are worth researching, if I have time. Just in time to shelter from the next deluge, we head underneath Moulsford Bridge.

The wooden footbridge takes us over the bog underneath the four-arch bridge. I perch on one of the railings as thunder cracks and lightning flashes alternate with the roar of trains directly overhead, and the rain buckets down. Fly is anxious to press on to the shelter of the woods ahead, but I suspect we're safer staying put for a while among the vegetation of the bog – comfrey, foxgloves and bulrushes. And I need to study the strangely twisting arrangement of the brickwork above and around us. The railway crosses the river here at an oblique angle, and the brickwork seems curiously braided, creating another Escher-like optical illusion.

Inconveniently, the rain persists and the dog finally has her way. Swathed in waterproofs, I step out into the storm, the dog scurrying ahead for the trees. Rain aside, the only thing that is beginning to dampen the spirits is the proliferation of signs along the path prohibiting all manner of innocent pleasures. 'No bathing. No picnicking. No mooring. No fishing', says one. Further on, there's a very unconvincing 'Bull in Field' sign which has been erected in what appears to be a garden of freshly mown grass and specimen trees – trees which surely would have been long trampled to a fibrous mush had there actually been a bull on the loose. Moreover, there is a curious lack of bullshit in this alleged field (apart from the words on the sign, perhaps), so maybe the bull wears nappies. Or else a member of staff, armed with a large shovel, is on permanent standby to clean up after the bull. Such signage has a tendency to convey a slightly mean-spirited message from landowner to the passing human traffic, which I can't help but interpret as: 'Bugger off, leave us alone and don't come back.'

Coming off the path to join a narrow lane towards Cholsey, a chap driving a Range Rover slows down to avoid drenching me in a puddle of water. A beige-coloured whippet sits on his lap, and another is curled up on the back seat. He winds the window down and asks with an almost toothless grin: 'Enjoying the weather?' We ask one another about our dogs, and Fly lies down, anticipating yet another idle conversation of some length. 'I've had whippets for sixty-nine years,' he says. 'They are the best dogs, good hunters.'

'What do they hunt?' I ask.

'Vermin – rabbits and rats,' he replies. 'I don't believe any other creature should be hunted. Just vermin. Rabbits and rats.'

* * *

Cholsey station is on the main line, and has a track branching off to Wallingford, 2½ miles away. The branch line opened in 1866 and passenger services continued until 1959. Virtually all Wallingford's station buildings – with the prime exception of the stationmaster's house – were demolished in the 1960s. The site was redeveloped for housing and a Habitat warehouse.

Such redevelopment might have killed stone dead any talk of trying to reopen the line, but not for a small band of enthusiasts who got together to form the Cholsey and Wallingford Railway Preservation Society. After years of hard work and fund-raising, the CWR managed to reopen the abandoned line to run steam and diesel-powered trains on weekends and bank holidays from Easter through to Christmas.

Before setting off on my walk, I contacted the CWR to see if I could visit them along my way. Bob Harrington, one of the volunteers who keep the line going, wrote back to me that this would be fine and suggested that on reaching Cholsey station, I could take the footpath that runs alongside the railway almost all the way into Wallingford. I'd arranged to meet up with Bob and some of his colleagues the next day so that they could give me a guided tour of their new station at Wallingford.

Fly and I find Cholsey station easily enough, but soon manage to lose the footpath. We're on the wrong side of the tracks. We should be on the western flanks of the branch railway, heading for Cholsey church, where Agatha Christie is buried. But we're on the eastern side almost level with the school. It takes us half an hour to right ourselves. Over to the left, framed by trees and a railway bridge, are three of the giant cooling towers at Didcot power station. A little way north-west of Didcot are the Sinoden Hills, known locally as the Wittenham Clumps, where evidence of Bronze Age settlement has been found. After crossing the A4130 near the level crossing, we come into Winterbrook, Agatha Christie's home for many years, and drop down towards the Thames path again. The queen of crime writers and her second husband, the archaeologist Max Mallowan, lived at Winterbrook House just outside Wallingford, an imposing brick and flint property backing on to the Thames; and thought to be the model for Danemead, Miss Marple's house in the fictional village of St Mary Mead. Christie was a regular user of this branch railway, apparently. 'Wallingford was a nice place,' Christie wrote in her autobiography. 'It had a poor railway service and was therefore not at all the sort of place people came to either from Oxford or London.'

An elderly man, out for a stroll in the early evening sunshine and fresh air, doffs his hat at me and bends down to stroke the dog. 'Are you enjoying your walk then?' he asks the dog, as she leans into his left leg.

'She's enjoying your attention and doubtless looking forward to another paddle in the river,' I reply on the dog's behalf.

Saturday morning. I meet Bob Harrington, his wife Susan and another CWR volunteer, Colin Sadler, at the Hithercroft Road Industrial Estate in Wallingford. Here the 'new' Wallingford station – a jumble of makeshift huts and maintenance sheds, carriages and locomotives – sprawls across a 5-acre site. All three are members of the permanent way gang – responsible for the track work and general maintenance of the railway.

'Shame you weren't here last weekend,' says Bob, inviting me into the café carriage. 'We had the *City of Truro* on loan from the National Railway Museum. She came on a lorry, all 91 tons of her, the first engine to exceed 100mph. Fastest thing on the planet.' About 3,000 people turned up to see and take a ride on three carriages hauled by the record-breaking locomotive, although this time she was tootling along at a sedate 25mph – the maximum speed allowed on the branch line. 'It was the biggest event we've ever organised – a wonderful weekend. Mainly families – we don't get as many anoraks as they do at Didcot.'

The CWR has a small museum, a shop selling souvenirs, books and ice creams, a Victorian ticket office dispensing cardboard tickets and a spacious café housed in an 1895 Cambrian Railways carriage. The carriage was brought here from the Lambourn Downs, where it had been someone's home for forty-odd years. Further down the track stand two shunters, *Unicorn* and *Lion*, formerly used by Guinness at the company's Park Royal Brewery and now on permanent loan to the CWR. Another carriage, built in 1849, has beautiful weathered sycamore panelling. Although closed to passenger traffic in 1959, the branch line remained in use for freight – mainly barley – until the former maltings building was demolished in 1982.

Apart from doing maintenance work on the track, Bob doubles up as ticket inspector on open days. Susan takes care of the attractive gardens she's created at the station, helps the stationmaster with his typing and makes posters. Sometimes she takes on the job of 'second man' (or is it second person?) on the diesel shunters, responsible for applying and releasing the handbrake and watching for signals. Colin does all the carpentry. All the plants are donated by members or well-wishers, and Susan is particularly proud of the old pig troughs that she has recycled into plant containers for the platform.

The society has about 280 members, of which about 25 turn out regularly to help with the routine work and run special events. Its president, Sir William McAlpine, who has his own personal railway in the back garden of his home near Henley-on-Thames, donated the huts that house the museum, the souvenir shop and the loos. A renowned enthusiast and supporter of railways, Sir William is great-grandson of Sir Robert McAlpine, who led the construction of the Fort William to Mallaig line at the turn of the nineteenth century.

'Some people dream of resurrecting a passenger service here, and some local canvassing work is about to start to see whether there's sufficient demand to make the idea viable,' Bob tells me over a mug of tea in the Wallingford railway station café. 'Personally, I feel that if a passenger service were to be restored, it would fail now for the same reason it failed in 1959. Most people want to use their cars to get about.'

One of the next jobs they need to get on with is to replace the roof on the museum and shop huts. Longer term, they plan to replace the decaying huts with a new station building, as has been achieved recently by the Chinnor and Princes Risborough Railway, provided they can secure the necessary funding and planning permissions.

Then there's the never-ending programme of sleeper maintenance and replacement. Bob and Colin invite me to join them as they walk along the track to Cholsey. They need to inspect it for any signs of damage following *City of Truro*'s visit the previous weekend – and check the sleepers. Volunteers have replaced 600 of them since Christmas, and laid new concrete sleepers along a quarter of a mile section.

'It takes four men to remove a wooden sleeper and between six and eight to remove a concrete one,' says Bob. 'Back-breaking work. You can recycle your wooden sleepers sometimes simply by turning them over, drilling new holes to fix them to the chairs and they can last another five years.' There are three types of sleeper on the line – metal, wood and concrete.

While run by volunteers, the CWR has a wealth of professional expertise among its members and supporters to draw on. Bob is a retired engineer, and another member, Colin Young, drives high-speed trains for a living. Staff from HMI Railways carry out inspections once or twice a year to ensure the railway is safe. A special safety inspection had to be organised before approval was given for *City of Truro* – insured for £500,000 – to run on it. Walking on, Bob points out the level crossing over the A4130 Wallingford bypass, which is funded and managed by Network Rail. There is a curious-looking wooden cattle grid to try to keep livestock and other animals off the line – the first wooden one I've ever seen.

'We get foxes, barn owls and a tremendous number of red kites over here – perhaps we'll see one today. Look at the wild flowers on here. We have a population of cowslips along here – they're a favourite of Colin's. Whenever we come across them on maintenance and they need to be removed, we replant them elsewhere.' Within a few hundred yards of Wallingford station, Colin and Bob are already spotting damaged or rotting sleepers, and the odd chair (which secures the rail to the sleeper) that needs sorting out. The heatwave has also left its mark, causing one section of rail to buckle slightly, expanding the track. The gap between rails should be about 8mm, but expansion has caused the gap to be reduced to 1–2mm. There's another big job to add to the list – 60ft of rail weighs a ton. Fortunately, two of the active CWR members, Tom and Dave Buckingham, work at nearby Manor Farm and their boss lets the society use his machinery for maintenance work. 'He's very sympathetic to the railway – which is a tremendous help,' explains Bob. When sleepers need replacing or re-siting, individual sleeper beds are dug out by hand with shovels. A mechanical digger is used for long sections of replacement work.

We're going over Winterbrook Crossing – Cox Farm is over the way. There are steel sleepers and chairs dating from 1931 on this section. Many of the sleepers come from the main line. Some of the chairs that support the rails date back almost to the First World War. What distinguishes them from more modern ones, I discover, is that they have two holes in them for the bolts, as opposed to three in the later versions.

Fly picks her way along the sleepers with us, occasionally pausing to suss out the trail of a fox or deer, sometimes running ahead or rounding us up from the rear. Having thick pads on her paws, she doesn't seem to mind walking on the ballast. We pass a P-Way Hut – a permanent way hut made of red brick. Gangers and layers put all their equipment in there. In the distance, beyond Streatley Hill and the Goring Gap, the Chiltern escarpment stretches right across the skyline.

We join the footpath I walked on the previous day and thread our way through St Mary's churchyard, Cholsey. 'Would you like to see Agatha Christie's grave?' Bob asks. He and Colin lead the way. She was buried here on 17 January 1976, her married name, Agatha Mary Mallowan DBE, carved on the headstone. A couplet from Edmund Spenser's 'The Faerie Queen' is engraved on the stone:

> Sleep after toyle, port after stormie seas,
> Ease after war, death after life, does greatly please.

We can hear the sounds of children and music playing at the Cholsey school summer fête, where a Puma helicopter from RAF Benson is one of the day's attractions. Soon we're looking at the cantilevered canopy of the Cholsey station roof. Colin reckons he can build something like this at Wallingford to replace the rotting huts – subject to planning permission and funding, of course. The group organises plant sales and concerts – well supported in and beyond the two towns – anything to raise money for improving on what they've already got and can offer to the public.

On the way back to Wallingford in Bob's car, he and Colin point out the stationmaster's house in each place – and a red kite wheeling overhead. Bob and Susan invite me back home as they've some photographs they want to show me. There's one of Bob dressed in his smart ticket collector's outfit in the ticket office, another showing the permanent way gang having lunch in the Toad (a brake van). Apparently, the GWR called all their working vehicles after fishy or aquatic things. There's a ballast wagon called a dogfish.

Over a sandwich lunch, I ask them why they lavish so much of their free time and hard labour on an old railway line. 'I enjoy men's company, having worked with men all my life. We have a laugh and enjoy meeting people. It gives us a bit of a purpose,' replies Bob. 'If people can see their children, or grandchildren, enjoying themselves, they enjoy the day too.'

Susan adds: 'I enjoy all aspects of what I do for the railway. It's a great source of interest and amusement. When the visitors tell us what a nice day they've had here, we all feel like we've really achieved something.'

* * *

Bob gives the dog and me a lift to the start of our afternoon walk to Didcot, dropping us off at a stile near Hithercroft Farm, a mile or so west of Wallingford. He asks about my intended route over the coming few days: 'I believe that Brunel and the others from the GWR had some of their meetings at the North Star [pub] in Steventon, the other side of Didcot. It was the halfway point along the Bristol to London route. You might want to call in there.' I thank him for a great morning and we wish each other farewell.

From Wallingford, I could have continued along the Thames path looping past Dorchester to approach Didcot from the north. But as I've walked quite a few miles along the Thames already, it seemed a better idea to go cross-country following the footpaths over farmland. About halfway between Wallingford and Didcot, the distant chuntering of a steam locomotive and short sharp blast of a whistle set the dog's ears waggling. It's the first of two Midsummer Steam Days at Didcot, and we'll be visiting tomorrow to catch the second one.

Soon we're passing the Didcot Community Wood, a collection of 7,000 trees and 2,000 shrubs planted by local people to commemorate the millennium, and maturing nicely. On our left, a new cycle path strikes off towards Wantage, meandering along the route of the disused railway line from Didcot.

* * *

Compared with the better known and altogether bigger Didcot Railway Centre, the Cholsey and Wallingford Railway could be regarded as a bit of a 'rag, tag and bobtail outfit', as Bob Harrington himself had put it. However, it does have one crucial advantage over its larger neighbour – namely 2½ miles of track linking town with village. Didcot has two lengths of track, the longest being a shade under half a mile, linking only one part of the centre with another. Equally, the Didcot Centre is entirely 'rail-locked', having no access other than via the pedestrian tunnel underneath the main line from the town's railway station, and the rails themselves.

This makes it a little less susceptible to the kind of vandalism that the CWR has experienced from time to time. The downside of the Centre's lack of road access is that signalling equipment, bits of engines, gas bottles and other kit have to be hauled upstairs from the pedestrian tunnel unless it can be shifted to the site by rail wagon.

'We don't have a huge number of serving or former railway staff among our members. I suppose if they're working all week on a railway, they want to spend their spare time doing other things,' John Minchin, Didcot's duty manager, tells me. On the second of the Midsummer Steam Days, John's doing a stint fielding enquiries from public and staff, as well as from me. 'I mean, I've got no railway or engineering background, although I grew up in a Bristol vicarage overlooking the main GWR line. When I started working here as a volunteer, I was running a record shop.'

There are about twenty volunteers on duty today, running three signal-boxes and three engines in steam: *Firefly*, *Earl of Bathurst* and the nameless 1338, built for the Cardiff railway in 1898. Now semi-retired, John has served as a volunteer at Didcot Railway Centre for most of his adult life. He began working there in 1967 after British

Rail withdrew from part of the 16-acre banana-shaped site, enabling the Great Western Society to move in with three rescued locomotives and some carriages.

Formed in 1961 on the initiative of a group of schoolboys, the GWS helped to pioneer the embryonic railway preservation movement that gathered pace all over the country in the aftermath of Beeching-inspired branch-line closures. Across the land, the proliferation of diesel-powered trains in the 1960s had either rendered redundant much-loved Victorian and Edwardian steam locomotives or at least put their future in jeopardy. Apart from a select few that British Rail promised to put on public display in its museums, most were destined to be broken up for scrap. The race was on to buy up the doomed locomotives before they vanished for good and, of course, to find somewhere to put them.

Public donations for such purchases began to flow in after a letter appeared in the *Railway Magazine* suggesting these unwanted engines ought to be preserved for posterity. Thanks to the voluntary effort of people like John Minchin, many were plucked from the scrapyards and given good retirement homes for future generations to maintain, enjoy and use. Twice a week, John still travels up from his home in Southall, west London to put in a day's unpaid labour of love, helping to rescue, restore and maintain old locomotives and coaches for the rest of us to admire and enjoy now. 'Most of ours were bought in the mid-1970s. Although we were paying scrap metal prices, we were still having to find £8,000 to £10,000 a time on average,' John recalls. 'The carriages were mostly bought off British Rail. The pure Great Western coaches had been downgraded to maintenance use by then. Some items we bought at auction or from museums.'

In the mid-1980s, Didcot was getting around a million visitors a year, although the number has since fallen to about 300,000 to 400,000. Perhaps this is not surprising as there are now so many preserved railways and heritage railway centres around the British Isles, more than 150 of them at the last count, for people to choose from.

The dog and I enjoy ambling around the site, and a short ride up and down the track on a carriage hauled by *Earl of Bathurst*. Down the platform, children and adults are queuing up to take a ride on *Firefly*, running on the broad-gauge track. The engine is actually a brand new one, built to the class specification of the original 1840 type. There are plenty of viewing points for the large number of men, weighed down by cameras and lenses, taking photographs.

The Centre isn't getting as many youngsters visiting as it used to, and the young volunteers team is only about twenty-strong. 'The rising generation of volunteer staff will have had no direct experience of steam engines in general use on the railways,' observes John, ruefully. 'They will have had no personal memories of steam in general operation, which is a shame.'

That hasn't deterred one youngster, at least, who calls in at the inquiry office, clutching a completed application to join the band of volunteers. Nineteen-year-old Nathaniel from Witney is keen as mustard to get started. He wants to train to become a stationmaster, and today he's on the roster to help out with general platform duties. Scanning the form, John tells the new recruit, 'Pleased to see you've done some first-aid training. That'll be very useful.'

'My dad's a big fan of steam, and I've always loved it,' Nathaniel tells me, when I catch up with him later. 'I used to help out at a railway at Gloucester, but I decided it was too far to travel. It's cheaper for me to come to Didcot and work here.' He's going to work alternate Sundays as he doesn't want to give up going to church. 'Yeah, I'm enjoying it so far – the people are really nice.'

As far as John is concerned, involvement in railway preservation has been a richly satisfying experience. He's worked his way up the grades to become an engine driver and taken part in several rescue operations. 'Getting locomotive 3822 out of Barry scrapyard, for example, helping to restore and overhaul the engine and putting it back into use for people to enjoy again – it's magic, a real delight.'

The beauty of the Didcot operation is that people of all ages can get involved, skilled or unskilled, and take their pick of a huge range of practical tasks, from track-laying, carriage restoration and engineering work, to catering or staffing the admissions office. Training is provided – enthusiasm is perhaps the main quality required of volunteers. As an independent charity, the GWS relies on subscriptions from its 5,000 members and events for its income. Hiring a steam train for the day is a popular choice for organisers of corporate hospitality, and for other special occasions.

Neil Sherry, an accountant, soon takes over from John in the inquiry office. *Firefly* has unfortunately developed a mechanical fault. One of the stays in the firebox has broken. Neil has to announce over the public address system that the locomotive will be out of action for the rest of the day. 'My grandfather worked on railway-owned ferries operating across the Solent, and one of my earliest memories is of him taking me to a locomotive works. I was smitten,' he says.

Not being mechanically minded, and having no desire to drive an engine, Neil helps out mainly on the administrative side, but sometimes does track work: 'I've always been interested in railways. When I moved to the Didcot area fifteen years ago, I decided to become a volunteer. It was time to put something back into a hobby that's sustained me all my life. What's satisfying is seeing the results of everyone's hard work, and it's nice to do something quite different from the day job – character building!'

Neil asks me what route I'm taking from Didcot the next day, so I show him the line of yellow highlighted footpaths and bridleways I'd drawn on the map as far as Wantage.

'I thought I'd go on the Sustrans cycle path to Upton, then over to Chilton and then climb up on to the Ridgeway,' I say. Like Bob Harrington, he suggests I might go further north instead, to take in Steventon. I decide to take their advice, particularly since the heat-wave shows no sign of letting up and there's not a lot of tree cover to shade us up on the Ridgeway.

* * *

Having been put up for two nights by my friends in Eynsham, Kathy Harvey and Dan Levy, I'm sitting at Oxford station the next morning waiting for a train back to Didcot to continue on my walk. Next to me is a chap with a bike waiting for the fast train to Paddington. He gives

Fly a stroke and, clocking the backpack, asks where we're off to. I give my usual spiel, tell him how far I've got and list some of the main towns I'll pass through en route to Penzance.

'I live at Sampford Peverell – backing on to the canal.'

'Well, maybe I'll go past your house then,' I said. 'I'll look out for you on your bike.'

Having struck off on our way in the morning sunshine – already strong – from Didcot station, we head west along the main road and then turn north on to the start of the Hanson Way, a network of traffic-free paths for walkers and cyclists linking Didcot, Abingdon and Oxford. The dog and I quicken our pace towards the shade of the trees and thick bramble bushes on either side of the path. They afford the added benefits of blocking out some of the traffic noise and exhaust fumes too.

The distinctive Sinedon Hills are visible occasionally through the bushes. We turn the corner from the eastern side of Didcot power station to walk between its rear boundary fence and a vast landfill site stretching up to the northern horizon. A huge flock of seagulls overhead eyes up the spoil below, freshly churned by two mechanical diggers. Next to the rubbish tip are gravel pits and a pair of lorries on the move. Flourishes of teasel, ragwort, nettles and flowering elder intersperse the ubiquitous clouds of cow parsley edging the meandering path.

Watching the white plumes pour from the lips of the cooling towers and drift into the blue sky exerts a dangerously hypnotic effect. Where do the clouds and the vapour meet? It's not easy to distinguish the two, looking directly above me, and trying to blot out the sun with a raised hand. The dog and I collide and I hit the ground with a thud.

3

Aunt Sally's Secrets Revealed
Didcot to Swindon

In which we find the North Star Inn, fall out under White Horse Hill, meet a pupil of John Betjeman's; and see Swindon at its best, having walked down from the Ridgeway, and come into town via Coate Water.

Through the perimeter fence, I see water cascading down an inner wall of the most northerly cooling tower. The unpleasant whiff wafting my way from the landfill site, exacerbated by the morning sunshine, pursues us over the footbridge spanning the brook into a field of broad beans. The pods are nearly the length of my hand, and a mass of white-flowering bindweed laces its way between the stems. It has taken us about half an hour to reach the north-west corner of the power station boundary – the site covers about 2 square miles. *Country Life* magazine surveyed its readers a few years ago to find out Britain's worst blot on the landscape. Didcot power station's cooling towers and chimney were ranked third behind Birmingham New Street railway station and, topping the poll, wind farms.

Instead of having to walk by road into Milton, as I had anticipated, we are able to take a pleasant short cut signposted 'conservation walk' through a glorious poppy field sandwiched between a school and an industrial estate, and then past a crop of green barley. Owing to an attack of vertigo halfway up the steps, I was unable to manage the footbridge over the A34, so we had to come down again and look out for the footpath running alongside the A road and the underpass a few hundred yards further on.

Since leaving Paddington, all the footpaths and stiles, gates and paths signing on my route to date have been above expectation and often quite superb. We enter Steventon near the old railway station site close to the Jehovah's Witnesses' building. The station has long been demolished and replaced by modern offices, but the two Brunel houses – as they're known locally – still stand a short distance apart. Both Station House, Grade II listed and now divided into two homes, and Brook House are thought to have been designed by Brunel himself. Station House, built in 1839 of ashlar limestone, with a gabled slate roof, served as the GWR's headquarters for a short time, the village being midway between London and Bristol. Between July 1842 and January 1843, the GWR board met here.

Brook House was built as a hotel to accommodate passengers to and from Oxford. While main-line services from Paddington had begun in 1840, there was no branch railway linking Didcot with Oxford for another four years. Meanwhile, passengers for Oxford had to travel by coach from Steventon. They would have done so for a lot longer had some of the local landowners and authorities had their way.

From 1837 to 1840 leading landowners, the University of Oxford and the city council opposed moves to build a rail connection with Didcot. By 1843, when a new bill was promoted through Parliament, most of the objections had been withdrawn, although the city council remained firmly opposed to the idea. The line opened on 12 June 1844, and the fares to London were 15s and 10s, compared with 5s by coach.

Fly and I pick a route through Steventon along the well-shaded cobbled causeway that skirts the large village green and well-tended allotments. At the North Star Inn, which Bob Harrington had suggested for a visit, I poke my nose round the door and ask a cleaning lady: 'Is it OK to bring my dog in? She's well behaved. On the whole.'

'There's a lady 'ere asking if she can bring 'er well-behaved dog in,' shouts the cleaning lady over her shoulder to the barmaid.

'She can,' the barmaid tells the cleaning lady.

'Come in,' says the cleaning lady, stretching out a hand. 'Animals welcome.' She brings out a bowl of water for Fly while I watch my half pint of Thoroughbred bitter being drawn straight from the barrel.

Dating back to the sixteenth century and originally called the Half Way Inn, it was renamed after the legendary locomotive that hauled the first train from Paddington to Steventon. It's a gem of a pub. Low beamed ceilings, high-backed settles arranged around a log fire, flagstone floors and a glorious absence of TV or jukebox. Paintings of locomotives cover the walls. There is no bar or counter, as such. Customers are handed their drinks through a door hatch from the taproom. The barmaid tells me a little about the history of the pub and the village, and hands me a copy of *The Story of Steventon* (1994) by A.L.H. Baylis. The dog enjoys a snooze on the cool flags, while I sip my beer and leaf through the book.

In the garden, there appears to be equipment for a game I can't immediately identify. There's a little white ball, shaped more like a light bulb than a ball really, sitting in a cup on a stick sunk into the ground. Behind it is a wall covered in black material with a white circle painted on it. I wonder what that's all about?

We set off again on the footpath heading south-west out of Steventon, over the level crossing over the main railway line, following East Hendred Brook to our left. Soon we hit a field with a herd of curious young male cattle heading straight for us. It's taken me years to suss out the best way of handling encounters with cattle, and a great deal of advice from real country people, as opposed to transplanted townies like me. What you have to do is keep walking, avoid eye contact with the cattle, and make yourself look big and fearsome (not always easy) by stretching your arms out each side. If that doesn't convince them to retreat, or just ignore you, you then use your voice. 'GET OUT OF THE WAY. GO ON. SHOOOO . . .!', for example.

What you should never do is start running, because there is a good chance the cattle will think this is a great game, or at least the most interesting thing that has happened all day. The point is they may be able to run faster than you and then gather around the very stile or gate you are aiming for to exit the field. This particular herd of cattle are too busy eating to take much notice of us.

Once again, it is a swelteringly hot day. Fly cools off in the brook and wants me to throw some sticks. There's a ford here and a nice bit of fast-flowing water. The dog finds a stick, picks it up and looks up at me with imploring eyes. I say 'Give', take the stick out of her mouth, throw it, tell her 'Well done, Fly' for bringing it back, and then the whole procedure starts afresh. We have to do this about 100 times a day on average. If she can't find sticks for us to play with, she will pick up twigs or dried up stems of maize or cow parsley instead.

The silage bales all around are covered in white plastic. We pass more broad beans and poppies. We get lost for half an hour or so, having strayed from a footpath. The sounds of the distant trains on the railway help to reorientate me back on to the right track. We are climbing fairly steeply now to the main road north of West Hendred – the A road between Didcot and Wantage.

Crossing the road to pick up the bridleway continuing opposite, we are on the cusp of two spectacular and very distinctive landscapes. Behind me to the north is a panoramic view of the vast Vale of the White Horse, and on the eastern horizon plumes of white smoke from the Didcot cooling towers rise into the sky. Ahead of me, to the south, are the Berkshire Downs and the foothills of the Ridgeway, which we will walk on later. It was simply too hot to walk on the Ridgeway today, there being so little shade up there. So we're turning right here, after crossing the road, to get on to the footpath towards Ardington, past more fields of broad beans and red poppies.

I pause to admire the grounds of Ardington House running down to the river. There are stacks of chairs on the lawns, as though there's been a concert over the weekend. Hearing the village clock strike 4 p.m., I realise we need to get our skates on if we're to reach our overnight B&B at the Old Vicarage, Letcombe Regis, in 2 hours time, as arranged.

The trouble is the scenery and hamlets all the way along here are real feasts for the eyes. The sun's shining, the dog's happy, and it's hard to take any sort of schedule seriously in such fortunate circumstances. The Ardington and Lockinge Community Woodland and Millennium Sundial, for example, cannot be ignored as so much work has gone into its creation. Particularly as the sundial has been arranged on such a grand scale, with thirteen pairs of standing stones arranged in a semicircle. You have to walk the well-mown paths, through the trees, between the stones to find the shafts of sunlight, experience its effects, and work out what time it might be.

It's getting close to 6 p.m. by the time we reach Wantage, where – near the town centre – we strike off along the footpaths to Letcombe Regis. Soon we reach the Greyhound Inn and opposite it, the Old Vicarage. Jillie Barton is used to welcoming weary long-distance walkers to her home, and tells me she recently put up an 82-year-old lady walking from John O'Groats to Land's End.

In the garden of the Greyhound Inn, enjoying the evening sunshine, a few people are playing the game I saw at the North Star Inn. 'What's it called?' I ask one.

'Aunt Sally,' a man replies. 'Each player gets six sticks to throw at the dolly – that's the white ball over there. The winner is the one who knocks it off the most times.' It seems to require much shouting, whooping and beer to play the game successfully. The sticks look a bit like skinny rounders bats. 'It's a traditional Oxfordshire pub game.'

* * *

I've managed to walk off with Jillie Barton's front door key, so she's driven up to meet me at the start of the footpath back into Wantage from Letcombe Regis to collect it. She looked after us really well in her beautiful house.

In Wantage, I need to call in at the bank, get something to eat for lunch and buy a torch for tonight as I am camping at Britchcombe Farm. John Betjeman lived with his family at The Mead, Wantage from the 1950s up to 1970. Before moving here, they lived at Uffington for thirteen years. We pass Betjeman Court, 'retirement apartments for sale', which look well designed and solid, and take a quick peek at the Betjeman Millennium Wood but we can't go in as dogs aren't allowed.

At the Vale and Downland Museum and Visitor Centre in Church Street (where Fly is made most welcome), the dog and I park ourselves in a mock-up of a Wantage tramway car in which, thanks to the wonders of touch-screen technology, we are 'transported' back in time along its rural route out of town to the GWR main line.

This is described as Britain's first steam-powered tramway, developed to provide a more efficient transport link between Wantage and the Great Western Railway. The nearest railway station was some 2½ miles north of Wantage. In 1875, the tramway opened for passenger and freight traffic. At first the tramcars were horse-drawn, but a year later the horses began to be replaced by the first ever steam-driven tramcar. The tramway offices and waiting room, built in 1904, now housing an estate agents, can still be seen in Mill Street. The tramway continued to carry passengers – about 36,000 a year – and freight until 1926, with freight-only services running until after the Second World War. It crossed the Wilts and Berks Canal on a custom-made iron bridge.

A statue of Alfred the Great, its most famous son, dominates the centre of Wantage, which has a nice bustling atmosphere. Lord Wantage, holder of the Victoria Cross, a noted social reformer and generous benefactor to the town, is said to have modelled for the statue. A great fan of King Alfred's, the peer paid most of the cost of commissioning the statue, sculpted by one Count Gleichen of Hohenlohe-Feodore, a nephew of Queen Victoria. Edward, Prince of Wales and Princess Alexandra unveiled the finished work on 14 July 1877. At Woolworth's I buy my torch, and ask the assistant where's a good place to buy sandwiches. 'Well, there's Waitrose, just further along the market place or there's a baker's opposite.' I opt for the baker's.

Although I had thought of climbing up on to the Ridgeway to get us to Uffington, again the heat of the day is too strong. We need to find a route where the dog can easily find water, so we pick our way through Wantage to join a well-shaded restored section of the Wilts and Berks Canal running between Stockham Farm, on the north-western outskirts of the town, and East Challow.

A family of moorhens flits between a couple of supermarket trolleys dumped in the canal. Soon we turn off the waterway to begin a steady ascent to the bridleway leading to Childrey. Once on it, fantastic panoramic views of the Vale of the White Horse open up to the north. It's quite thickly wooded here between the villages and hamlets. We pass a farm worker spraying a field of wheat and hurry on to Childrey nestling in a little dip in the land. Fortunately the wind is blowing away from us.

From Childrey, we take the minor road on to Sparsholt, passing the sign to Challow railway station. Hay bales have been neatly stacked into an elongated pyramid in the field to my right. At the Hatchet pub in Sparsholt, the landlord fills in yet more gaps in my knowledge about the Aunt Sally game. 'There's a chap in the Oxfordshire league who's so good he can knock down the dolly with four out of six sticks whilst blindfold.'

By the time I climb up the hill to Kingston Lisle, the wind is getting up and I don't like the look of the black clouds gathering above me. I've cut my hand on a bit of bramble and the dog's sulking because she's not had any sticks to chase for at least half an hour. I turn round to look back at the view, and in the distance I can make out four of the cooling towers and beyond them the Chiltern escarpment. We're both a bit hot and bothered by the time we get to Kingston Lisle, so we call in at the Blowing Stone Inn for more refreshment. Fly is served her water in a champagne bucket. They're very civilised in Kingston Lisle.

The village has a beautiful collection of brick and flint cottages, and horse racing stables. We pass a couple of thatchers in the process of stripping off some of the old thatch from a cottage near Fawler, a hamlet to the north-west of the village. The quickest way into Uffington would be to carry on along the minor road, but there's quite a bit of traffic about, so we turn off on to a footpath near Fawler Farm.

Soon I can see the tower of Uffington church, White Horse Hill and the Dragon's Hill, the latter being an odd-looking chalk mound shaped like an upturned cupcake. Legend has it that St George slew the dragon on this hill to save the king's daughter from being sacrificed.

I've arranged to meet my friends Mark and Tanya Hughes at Britchcombe Farm campsite, as they're kindly lending me their tent for the night. All the B&B accommodation in the area that I'd contacted was either booked up or didn't take dogs, although one lady did offer to quarter Fly in her garage for the night. Fearing that a solitary night in a strange garage might test Fly's loyal support to the very limit, I decided that camping might be preferable.

Unidentifiable bits of the Uffington White Horse are revealed to me through leaves and branches of the trees lining the footpath. A tail, a head – or is it a leg? There's just time to walk into Uffington before meeting up with Mark and Tanya. Underneath a giant blue, white and red 'Vote Conservative' placard lashed to a tree, I leave the footpath and step on to the road leading from White Horse Hill into the village. The general election was nearly two months ago, but such exhor-tations – rather like the green, white and red 'Keep Hunting' signs – seem to have become permanent fixtures.

More black clouds gather overhead and gusts of wind are blowing through the trees. At the friendly and well-stocked post office and stores, I ask the assistant whether anything remains of the long-defunct Uffington railway station. 'A few station buildings – now private residences, I think,' she replies. 'If you want to go have a look, there's a path up to the footbridge over the railway line – or else you can walk up the Baulking Road.'

I tell the assistant and one of the other customers in the queue that I'm booked into the campsite at Britchcombe Farm for the night and feeling a little queasy about the forecast thunderstorms. 'You might be able to get B&B at Fenella's,' says the other customer.

'Except she's gone off to Portsmouth for the Battle of Trafalgar re-enactment,' interjects the assistant. 'It's the bicentenary. Her husband's very boaty.'

As the rain clouds descend, I take a quick walk round the village, almost breaking into a run, past the church with the octagonal-shaped tower and the Tom Brown's School Museum. Walking back along the road up to the farm, the Uffington White Horse looks more like a squiggly long division symbol. I saw it from the air a few years ago. They were doing helicopter rides from the Uffington Show on August Bank Holiday weekend. From above it actually looks like a galloping horse and I couldn't help thinking then – as now – that it was perhaps designed to be seen from the air. Unlikely, I suppose, given that it was carved on to the hillside at least 3,000 years ago. Who Knows?

On reaching the campsite, I find that Mark has already erected the tent. 'Wow – this is a palace!' I tell him, trying to suppress my growing sense of gloom. 'We'll be nice and snug in here.' I knock at the farmhouse door to report my arrival and pay Marcella Seymour the three quid camping fee.

A quick visit to the Fox and Groom at Uffington is in order before Mark and Tanya abandon me to my fate. Enveloped by intensifying gloom, the drizzle now turning to heavy rain, we huddle with our dogs, drinks and crisps underneath a sunshade in the pub garden listening to the distant thunder.

Bang on cue, the first flashes of lightning illuminate the campsite seconds after Mark and Tanya drive off back to Malmesbury. Swiftly shooing the dog inside the tent and zipping up the entrance flap, we listen to the rain crashing down on the tent like a volley of cannonballs. Can the re-enactment of the Battle of Trafalgar really be taking place as far away as Portsmouth? Fly eyes me accusingly, before delivering the ultimate snub – curling up to go to sleep with her back to me. 'Why on earth have you brought me to this hell-hole?' is the unspoken question hanging in the air. The evening's entertainment consists of watching puddles form through the plastic windows above me, and guessing how long they'll take to turn into rivulets; and watching the dog catching the bluebottles imprisoned with us in the tent.

* * *

Even the cattle join in the dawn chorus, which wakes me at about 4 a.m. Something warm and hairy is pressed against my cheek – thankfully not a wild animal that's broken into the tent, just the dog's rear end. 'Did you get much sleep?' Marcella asks me as I struggle to dismantle the tent.

'Oh yes, several hours,' I lie.

She invites me to call in at the farmhouse once I've finished packing up to sign her visitors' book.

Marcella Seymour, a tall, sprightly 78-year-old, took charge of the 600-acre farm and small campsite after her husband died. Before he passed on, she was his 'gofer' she says. Before she married, her father had run the farm. Some of her ewes graze the lower contours of the Manger, the spectacular dry valley created by a melting glacier at the foot of White Horse Hill. On Sundays and Bank Holidays she and her helpers serve cream teas to tourists and visitors seated at picnic tables, the resident hens and roosters pecking at the crumbs.

Sheltered by woods on the lower slopes of the hill, Britchcombe Farm dates back to 1649. The old stables and visiting grooms' accommodation, little more than a shed called the cot, now house the campsite's toilets, shower block, and the kitchen where the cream teas are prepared. Recent campsite visitors included a group of Woodcraft Folk, who made nightlights out of tin cans with holes they'd pierced to let the light through.

Born in Uffington, Marcella remembers John Betjeman well. One of her abiding memories is of him trying to teach her how to recite poetry in the mid-1930s, and on one occasion, Betjeman throwing a copy of *Tom Brown's Schooldays* at her. 'He was trying to teach me to recite poetry – in Berkshire dialect. It was a poem quoted in *Tom Brown's Schooldays*, about the Berkshire Pig. There was the word "cussed" and I wouldn't say it cos I was always told that it was a swear word, so I said "awful". He threw the book at me and said, "Never heard anything so foolish in all my life."

'Another time we went to an event at Ashbury. They were putting something on – he and I were reciting poetry there, and I remember him producing a pig's head at one point. Later he brought an Irish friend of his up here one night for dinner – who subsequently got killed in Persia, just before the war. My mum and dad did bed and breakfast in those days, and he used to bring people up here.'

In recent years, Marcella has attended many meetings with people from the National Trust, English Nature and various other bodies concerned with the conservation of White Horse Hill and the surrounding landscapes. 'Some of 'em wants the grass three foot high, some of 'em wants it short for the wild flowers. What I worry about is my ewes, 'cos some of them's been killed by dogs, with the public going over the hill. Some of them gets real abusive when I asked them to put the dog on its lead.'

* * *

Heading up towards the crossroads, we turn left towards the White Horse and Uffington Castle, an Iron Age hill fort. Much of the Vale is concealed under a damp mist. A couple of tractors and a National Trust ranger's vehicle pass us on the track before we step on to the Ridgeway towards Wayland's Smithy. Systematic conservation and maintenance of the White Horse, cleaning, repairing and weeding, are known to have been carried out at least since the early eighteenth century. A fictional account of a cleaning 'festival' during the Victorian era, involving games, performing acrobats and feasting, was given by Thomas Hughes in his novel *The Scouring of the White Horse*. Following the First World War, complaints about the neglected state of the horse reached the national press. 'The famous White Horse on the Berkshire Downs near Uffington, to which reference is made in *Tom Brown's Schooldays*, is but a shadow of its former glory,' reported the *Daily Mail* of 26 August 1922. 'It can scarce be seen for want of scouring. It has become so dirty of late that travellers on the Great Western Railway line, who still remember it as it was many years ago, standing out clear and active on the long range of chalk hills, complain that something ought to be done to renovate it.'

Fly pads her way across a cattle grid and I unclip her lead. Looking back at the White Horse, the carving now appears as a series of fluidly drawn pen strokes. It's a really excellent section of track, with separate paths for walkers and cyclists to choose from. The dog is spoilt for choice of many milky white puddles to drink from. After ten days of trekking, the temperature is comfortable for the first time. The hedgerows are ablaze with the colours of elder, bramble, pink campion and bindweed in flower.

A 'No Entry' sign informs us that, under recently passed legislation, the Ridgeway is closed from 29 November to 30 April to motor traffic. For years, joyriders in 4 × 4 vehicles have turned many parts of the Ridgeway into dangerously rutted mud baths, so I say 'Hallelujah'. A photographer laden with equipment loiters at Wayland's Smithy, a neolithic chambered long barrow, encircled by a stand of beech trees, presumably waiting for the mist to clear.

By the time we reach the water trough, the sun's shining through white puffy clouds. The dog plonks her front paws on the top edge of the trough and laps up the murky liquid. A curious sign next to it says: 'Tap. Trough of water for animals. No washing.' A weasel – or perhaps it's a stoat – stands on its hind legs, a few feet away from us, sniffs the air and has a good rub at its whiskers. A goldfinch takes a bath in a puddle, as two or three buzzards swirl overhead.

A short distance from Bishopstone and Idstone, we step out of Oxfordshire and into Wiltshire. The path gradually descends now and away to the right Swindon's sprawling form begins to unravel across the north-western horizon. The spire of Christ Church, Old Town, pierces the blue sky, and, fanciful as it may be, I think I can make out the tower of St Sampson's Church, Cricklade in the far distance. Some free-ranging Gloucester Old Spots go about their daily business over to the right. Sows snooze in the shade of their metal shelters, a few others rummage about, while seagulls and crows flap above. The low rumble of traffic, getting louder, indicates we're approaching the M4.

Just before the Roman road down to Wanborough, the garden of the Shepherd's Rest pub looks too enticing simply to pass by. A couple of office workers, mobile phones clamped to their ears, are waiting for their lunch to arrive. After putting in my order, I flick through a copy of the *Daily Mail* and read the denunciations of the ex-Transport Secretary Stephen Byers by angry Railtrack shareholders.

Ahead of us, as we come off the bridge over the M4, stands Liddington Hill, crowned by a clump of beech trees and an Iron Age hill fort, surrounded by a shimmering haze of blue linseed coming into flower. Swindon's new Great Western Hospital, built on the south-eastern outskirts of town near Junction 15 of the M4, remains within sight for the next hour or so's walking.

After crossing the A346 to Marlborough, my route-finding around Chiseldon is hopeless. Chiseldon is on the cusp of different OS maps and I don't seem to have the right one with me. Someone is bound to be able to direct me on to Route 45 (of the National Cycle Network) up to Coate Water Country Park, Swindon's largest area of public open space.

Having found Route 45 and walked along it a few hundred yards into Chiseldon, I promptly lose it again in a modern housing estate. Despite flagging down several people out and about, I find that none knows the way. Eventually a girl aged about eight, about to get in a car with her mother, was able to direct me with some confidence. She'd recently been on the route to Coate Water on a school outing.

The dog and I trudge wearily past Chiseldon sewage works, and Hodson, finally joining Sustrans Route 45. Faced with the prospect of another vertigo-inducing climb on to a footbridge, fatigue overcomes fear. Straining at the leash, the dog drags me over the M4 in no time at all.

Once the traffic noise abates, the route up to Coate Water is tremendously green and calming, and provides one of the most pleasant approaches to Swindon. Sitting by the water near the café, the dog having scoffed the remains of my ice cream, I admire the fabulous rural views up to the Ridgeway and the downland – views likely to be partially obliterated by more new housing and a new university campus in a few years' time if current plans for developing 500 acres of farmland are approved.

It was in the early 1820s that the Wilts and Berks Canal Company built the reservoir of Coate Water to top up the canal as it passed through Swindon. After the Great Western Railway grabbed most of the canal's trade the waterway fell into decline. In 1914, the company sold Coate Water to Swindon Corporation for recreational and leisure use for the townspeople, principally boating, angling and swimming. An Art Deco diving board was constructed in 1936 out of 211 tons of concrete and almost 2 tons of reinforced steel, rising 32ft high. After a national polio scare in 1958, swimming and the use of the diving board was banned for good.

Already the 'save it, don't pave it' campaign to maintain Coate Water as a country park has attracted 20,000 signatures, and I add mine to the petition on display at the café.

* * *

My friend Alison Griffiths and her border collie Badger join us for the next leg of the walk. In fact she's kindly worked out the route we're going to take both into Swindon centre and back out again towards Lyneham. Alison has worked in the town for many years, first at the Princess Margaret Hospital, and since that closed, the new Great Western Hospital. We share an interest in old canals and railway lines and the route she's designed offers plenty to see on these themes. In order to avoid the busy Marlborough road, she takes us north-west through Coate Water Country Park and the attractive Coate Tree Collection. In memory of family members, locals themselves have planted many of the trees whose foliage, besides looking good, helps to block out some of the motorway noise. We cross the deserted recreation ground, emerge into a close of 1930s semis and bungalows, Downs View Road, and having crossed the Marlborough road, head for Old Town Railway Path.

This route once formed part of the former Midland and South Western Railway, which linked Swindon with Marlborough and Southampton. Lined by mature trees, the path and the sides of the cutting we're walking through are filled with dappled sunlight, and well used this afternoon by a mix of joggers, walkers and cyclists.

'Great path, Alison,' I say. 'I'd never have found it – how did you know it was here?'

'I used to cycle to work this way sometimes. Fell off my bike once in pitch darkness. Went right over the handlebars. Had to have an operation on my thumb.'

'Really? You broke your thumb?'

'No – pulled the ligament. The doctor said I'd got Poacher's Thumb.'

'You're kidding.'

'The doctor said the injury was similar to a condition that used to be common among people using guns a lot, pulling triggers. He called it Poacher's Thumb.'

'Well, I never. You learn something new every day.'

'It was excruciatingly painful. If I hadn't had an operation on it the next day, I could have lost the use of the thumb permanently.'

'Must be awful to lose the use of a thumb – such useful things. Is it all right now?'

'Absolutely fine.'

'Thank goodness for that. What's this over here?'

You can see different strata of rock on the cutting, and the information board tells us that we're walking through the Old Town Railway Cutting Site of Special Scientific Interest. The rocks and fossils revealed here, as a result of excavation for the cutting, were laid down when the shallow Jurassic seas covered the area, between 155 and 145 million years ago. The layers of yellow limestone around here were together called Cockly Bed, so named because it was found to contain large numbers of fossils of cockle-like shells. Some of the sandy limestone beds contained an abundance of glauconite, a greenish clay mineral.

Every so often along the path there are rusting bits of old railway infrastructure, and recently placed examples of rather surreal works of art, bearing cryptic messages. Engraved on the first one we see are the words: 'Our wheels relinquish and seize.'

Through the trees to our left, on the southern horizon, we get glimpses of the Wiltshire Downs, Liddington Castle and the old Wroughton airfield complex of former hangars. The Science Museum's 'large objects' collection is housed here. In the foreground, Alison points out the Intel building and Swindon's famous 'Front Garden'. This is the large area of trees and small fields that sweeps down from the hill on which the Old Town stands to the M4, providing a welcome rural buffer between the two. Again I'm savouring this view because it won't look this good for much longer. Work on building 4,500 new homes on the Front Garden is about to start shortly, although the developers have promised to keep about half of the land as open space.

After passing the site of Rushy Platt halt and what used to be Rushy Platt Farm (now a housing estate), we come across another strange piece of art. This time the legend on the stone wheel says: 'Stepping out of character, you interrogate a chaos of bearings.'

Badger lifts his leg and urinates on the work of art, and we move on without further ado.

Soon we're crossing the Wootton Bassett road near the Dick Lovett Garage, to walk across the recreation ground, past some meticulously kept allotments and the Unigate depot. Alison points out the Arkells Brewery in the distance. The bramble bushes are heavy with blackberries, still small and green.

On to the parish of New Swindon via the GWR Park. Although the bandstand has long gone, the old railway tradition of staging an annual children's fête here is maintained and we stop to read a poster advertising this year's event. Over the road stands St Mark's Church, designed by Sir Gilbert Scott for the GWR workers and their families, built by the company between 1843 and 1845 at a cost of £5,500. Scott also designed the vicarage and the church school.

Once through the railway village, Alison, the dogs and I catch a bus back to Coate Water, as we're anxious to fit in a visit to the Richard Jefferies Museum. It only opens twice a month and it would seem wise to strike while the iron's hot. Richard Jefferies (1848–87) was a journalist for the *North Wilts Herald*, whose boyhood exploring the countryside around his home at Coate Farmhouse, now the museum, inspired a passionate devotion to nature and nature writing. My mother once gave me a biography of him written by Henry Williamson, author of *Tarka the Otter*. Williamson was a friend of her father Huw Roberts, a signalman for the GWR around Fishguard.

'No, you're not too late – come in.' Chris Bowles, who helps run the Richard Jefferies Society and the museum established in his memory, gives us a warm welcome and waves us in. The society operates the museum, but Swindon Borough Council owns it. We're only the fifth set of visitors he's had all afternoon.

'It doesn't surprise me,' I reply. 'We didn't see a single sign from Coate Water.' Chris shows us round the house, pointing out the bedroom window, from which the author, as a child, would frequently escape at night, shinning down the drainpipe to roam around at will in the darkness. 'If you stay in a place too long, your heart gets dusty like a clock on a shelf,' wrote Jefferies in one of his works, which Chris quotes to us at the top of the staircase.

The society has about 400 members, some in Japan and America, where the author is better known than in the UK. 'We get people coming here from all over the world, China, India, America. Often they come to Swindon to train for their work, and they comment, without knowing much about the history of the area, that it's a town that has no soul, with the railway works having gone.'

'Do you think that's true?' I ask.

'I think it's changing. In the 1970s and '80s, it did feel different with the railway declining and the engineering works closing,' Chris replies, perhaps diplomatically. 'Now the town is re-creating itself, with more outsiders coming in, it is perhaps settling down again. Outsiders seem to take more interest in the town sometimes.'

4

Steamopolis

Swindon to Chippenham

In which ex-locomotive drivers and platemen, carpenters and guards, tell me about life at the Swindon works, we discover newly restored sections of canal near Wootton Bassett, lose a vital piece of kit and hear sobering news.

In 1841, Daniel Gooch, the GWR's chief locomotive superintendent and Brunel's right-hand man in many ways, recommended the level green fields a mile or two north of Swindon as the site for the company's principal locomotive repair workshops. What is now known as Old Town clung to a windy hillside on the northern outcrops of the Wiltshire downs, supporting a population of about 2,500. Few could have imagined how the sustained influx of railway workers and their families over the coming years would change their lives and rural surroundings. Whether they liked it or not, a new town would be built on their doorsteps – founded on the magical powers of steam.

The close proximity of two fully functioning canals, the North Wiltshire and the Wilts and Berks, made the site ideal for the transport of coal and building materials. Furthermore, a station was needed here anyway to provide a junction between the Paddington to Bristol line and the Cheltenham and Great Western Union line, also under construction around this time.

Within five years of the decision to site both the engineering works and a station at Swindon, the workshops were not only repairing engines but had built the company's first engine, the *Great Western*. The GWR company had to build New Swindon from scratch and by 1853 an estate of more than 240 workers' cottages had been laid out, linked to the locomotive works by a long tunnel underneath the railway. It was one of the first planned Victorian estates, other later examples being Bournville, Birmingham, built by George Cadbury, and the Lever Brothers' model village of Port Sunlight on Merseyside.

The works rapidly expanded to include a carriage and wagon works, management and administrative offices, a gasworks, general stores, a drawing office, laboratory, telephone exchange, laundry and fire station. By the early twentieth century, it had become one of the world's largest and best equipped railway engineering works, spread across 326 acres and employing 14,000 men and women. Workshop A alone, the principal erection shop where all the relevant component parts were brought together for assembly, covered a whopping 11½ acres. The locomotive department had the capacity to build two engines a week and repair more than 1,000 a year. It came to be known locally as 'inside'.

Workshop R now houses STEAM – formerly the Museum of the Great Western Railway – and other on-site conversions include a discount shopping centre and the National Monument Records Office. The slow regeneration of the old Swindon engineering works site has been reinforced recently by the completion of the new eco-friendly National Trust central offices, built in blue-grey Staffordshire engineering brick, accommodating more than 400 staff.

What I've always enjoyed in the last few years about my visits to the STEAM Museum in Swindon is the way it brings alive its social history. It's good to have the excuse to visit once again.

The artefacts of the GWR – everything from magnificent loco-motives down to the polished carriage doorknobs and china tea plates – are imaginatively and lovingly displayed. The clothing, posture and setting of the mannequins, the expressions on their faces, give a sometimes unnerving glimpse into workers' lives.

Alan Philpott, a third-generation railwayman and former works carpenter, shows me a painting depicting the area in 1849. On it I can see the emerging new town of Swindon's parish church of St Mark's, whose consecration in 1845 was attended by Brunel – his only recorded visit to the town after the opening of the works; the 'company houses' – now better known as the railway village; the park and bandstand; and the barracks accommodating about 100 single men. There's also a covered market, established after complaints from the railway workers' wives about both the high prices they had to pay in Old Town shops and the long trudge up the hill to reach them.

Thus began the slight tension between the gentrified, traditional old town and the rapidly expanding new settlement of railway workers and their families, one which some say never quite disappeared. Besides being built and repaired here, operating locomotives were changed, too. From Swindon eastwards, all the way to Paddington, the railway was almost completely level, thanks largely to Brunel's meticulous surveying. However, the gradients between Swindon and Bristol, on the southern and western fringes of the Cotswold Hills, were steeper, requiring locomotives better suited to hauling carriages and wagons up the steady inclines.

'It's such a shame the canal through Swindon was filled in during the 1960s,' says Alan. 'It would have made a wonderful water feature for people to enjoy now.' On the bottom left-hand corner of the 1849 painting, Alan points out another important Swindon landmark – derelict for many years – the GWR Mechanics Institution, originally providing outstanding facilities for recreation and entertainment, adult education and cultural activity. The workers themselves supplied many of the first books for the lending and reference libraries here. There were games rooms, a theatre and later showers. Actor Roger Livsey, the star of many celebrated Powell and Pressburger films in the 1940s, including *The Life and Times of Colonel Blimp* and *A Matter of Life and Death*, ran a repertory company at the Mechanics Institution after the Second World War.

In the early days, craftsmen walked from as far away as the Midlands and Wales to seek jobs at the Swindon works. There was no shortage of willing, skilled recruits, or youngsters seeking apprenticeships. Alan Philpott's own grandfather, brought up around traction engines, was taken on after walking over the downs from Devizes in the 1890s in search of work there.

An annual highlight for children like Alan, growing up in the 1930s and 1940s, was the Swindon works children's fête in the park. They had a funfair, sometimes a circus, acrobats and tightrope walkers, and fireworks at the end creating an image of the King and Queen in profile. 'You had to take your own cup with you for the tea you had with a big slab of fruit cake,' recalls Alan. 'With thousands of people descending on the park, they didn't have enough cups to go round. One of the firemen invented a cake-cutting machine to speed up the service!' As a child he also remembers using the railway pharmacy, where prescriptions were dispensed from behind a waiting room screen via a hatch in the wall.

A comprehensive range of health services was provided by the company for staff and dependants. A married man with children under sixteen in 1947 typically paid 8*d* a week into the GWR Medical Fund Society, and for this he and his family received any necessary hospital treatment, all dentistry, chiropody and optician's services, and prescriptions. Aneurin Bevan's infant National Health Service, set up in 1948, drew heavily on the ideas and practices pioneered in this railway town. When Alan began his apprenticeship in 1949, just after nationalisation of the railways, memories of war were still fresh: 'There had been guns on the works roofs during the war, and many of the workshops were used for building submarines, landing craft and bombs weighing 2,000lb and 4,000lb each. In the event, Swindon itself wasn't bombed much. I think the Germans wanted the engineering works too much to bomb it to any great extent.'

Alan spent nearly thirty years as a fitter, turner and erector at the Swindon works, staying on a couple of years after its closure to help clear the site ready for redevelopment. 'As long as you did your work competently and behaved yourself, you were fine. You had a job for life, cradle to grave welfare support and the opportunity for education.'

Just as rural workers relied on the village church clock to regulate the working day, the Swindon works had its hooters, audible up to 6 miles away. Punctuality was next to godliness. Under British Rail Engineering Ltd (BREL), as under the auspices of the GWR, the old steamship's siren was sounded 45 minutes, 10 minutes and 5 minutes before the start time. Fairly early on your route around the museum, you can see a re-created foreman's office where the boss is carpeting a worker for being late – thus putting at risk his whole future.

Although long retired, Alan returns every week to his old workplace, now as a volunteer, helping to pass on the story of the GWR, and Swindon's role in particular, to rising generations. Given the dwindling number of men and women with first-hand experience of driving or firing steam locomotives, or helping to build them, he and his workmates perform a vital task, and they do it with dedicated enthusiasm and bonhomie.

Alan passes me on to Peter Pragnell, the 'baby' of this band of ex-railwaymen, a mere seventy-year-old, who followed his mother and father into the Swindon works a year after Alan in 1950. On my tour of the museum, we've reached the storeroom. Catching my eye is a wooden-framed glass case containing the Brunel Collection of fifty-six different types of stone found in the districts served by the GWR, each about 4 or 5in square. Absolutely exquisite.

Soon, we're talking toilets, timekeeping and bowler hats. In the 1960s, one of Peter's jobs as a carpenter was making plywood partitions to cover up the original circular inspection holes in the doors of the WCs. 'The maximum time allowed for toilet breaks had been 10 minutes. If you exceeded that limit, even by a minute, you would get 15 minutes' pay docked. The holes in the doors enabled the supervisor to see who was where, and for how long. So sitting on the loo filling in your football pools coupon, as some did, was not without its risks!' he says.

Peter beckons me to another exhibit. 'Look at this wooden toilet seat. Guess what they used as templates to get the size of the hole right.'

'No idea. I dread to think,' I reply.

'Bowler hats,' says Peter, breaking into gales of laughter. They never had to look far for one. All the GWR head foremen wore bowler hats. He walks me over to view the Royal Coat of Arms made for Queen Victoria's funeral train, later dusted down, cleaned up and put to the same use for ensuing monarchs' funerals.

Soon we reach a mock-up of a railway carriage under construction, a wooden frame made of ash, suitably pliable for steaming the roof arches into shape, with teak and oak added for durability. 'Recognise him?' asks Peter, nodding at the life-sized mannequin of an aproned carpenter, bent over his lathe. He was asked to model for the figure a few years ago, a process that required a cast to be taken of his head – not an experience he would wish to repeat.

The Swindon works made virtually everything the railway needed, right down to the leather blinkers on the horses' harness. In an industry where accidents and injuries were common, the works even manufactured their own artificial limbs, although the company's motivation for doing so was not entirely altruistic. Losing a limb was no grounds for compensation or early retirement. You were simply fitted with a brand new wooden one, crafted on-site, and then despatched to join the invalid gang. In the long era of thrift, making do and mending that prevailed right up to, during and beyond the Second World War, the GWR tried to recycle almost everything that ended up as waste among the ash and wood shavings on the workshop floors. Fallen nuts and screws were swept up and given to the invalid gang to sort into piles of the correct type and sizes, ready to be returned to the storeroom. The working conditions were often hot and filthy, as well as hazardous, in those pre-Health and Safety Executive days, long before the phrase 'compensation culture' had been coined. Alan's boss died of asbestosis, as did several other colleagues of his acquaintance.

A vivid and damning eye-witness account of gruelling monotonous labour and appalling hardships endured by workers in the foundries, blast furnaces and blacksmith's shops that made up the complex was published in 1915. Alfred Williams based his book, *Life in a Railway Factory*, on his twenty-three years' experience working there and the degrading treatment of the men he saw meted out by the foremen and managers. The book, though well received elsewhere, was castigated in a GWR magazine review, which portrayed Williams as a mentally unstable 'nature lover', exhibiting much the same sneering cynicism towards his fellow workers as he did about the GWR management: 'At the forge Mr Williams was utterly out of his element,' the reviewer wrote, 'while at work his mind was roaming in the fields, on the hills, among the flowers of the countryside, or straying to the books of his beloved library. He used the tarred surface of his boiler for marking thereon extracts from Greek authors, which he was bent on mesmerising, and felt pain and mortification when his less classical mates poked fun at these exercises or put a fresh coat of tar on the boiler.'

* * *

Peter is ushering me past a display showing a pair of women riveters in denim dungarees, their hair tied back in brightly coloured headscarves. He's anxious for me to meet some octogenarian train drivers, fellow volunteers who turn out every week to show curious ignoramuses like me how things were done in the age of steam. They'll be off home at 3 p.m., so I need to step on it. I can linger later by the Wall of Names.

At the age of fifteen, Ted Abear began his apprenticeship at the Swindon works in 1946, and he became a fireman a year later. It took another fifteen years before he was deemed sufficiently experienced to land his first job driving a locomotive. Like Alan, he's a third-generation railwayman. In the late nineteenth century his grandfather worked on the line at Old Oak Common near Paddington, which Anne Boston and I passed a fortnight ago while walking the Grand Union Canal path.

Ted and I are standing aboard the cramped powerhouse of *King George V*, which in 1927 was famously dismantled into three sections for shipping to Baltimore for celebrations to mark the centenary of the Baltimore and Ohio Railway. Ted's showing me the vast hole, practically a cavern, into which the coal – preferably from the South Wales minefields – needed to be shovelled. 'The fireman keeps the engine going,' explains Ted patiently, possibly for the millionth time. 'The coal fuels the fire. The fire heats the boiler. The heated water creates the steam that drives the engines.' I try to take it all in.

Chucking shovelfuls of coal down the furnace in the right way at the right time may sound fairly mundane, but required the kind of precision timing, alertness and energy that might be beyond most of us today. The fireman's art was, physically and mentally, extra-ordinarily arduous. If you didn't do the job competently, the driver couldn't do his either. And it was usually a he. The reason being that no engine driver would be let loose on a locomotive unless he had served his time as a fireman. On a typical four-hour journey, the physical strength and stamina required of firemen – hauling up, swinging round and launching perhaps 8 tons of coal at the fire – tended to be beyond the capability of most women. So if women did manage to work their way up through the GWR grades, it was usually via the clerical, administrative and management ladder.

And women who married into railway life often had a hard time of it too, struggling to bring up their children alone for long periods. The drivers' and firemen's rosters dictated their lives for weeks and often months ahead. Ted recalls doing a forty-eight-week 'link' or roster, mapping out his entire life for almost a year, including 'double home' jobs – overnight periods spent away from base. These schedules were known in railway argot as 'washing machine or telly links' – by the end of it they'd earned enough to buy a Servis twin tub or a black and white television set. No one had heard of paternity leave, so when Ted's daughter was born, he had to carry on with his duties: 'You could only get a day off if there was a death in the family,' Ted recalls.

Ted's service ended in 1987, when he joined the Friends of the Swindon Railway Museum, then located off the Faringdon Road. He's now vice-chairman of the Friends. As well as turning out every other Saturday to answer questions from tourists and visitors, many of the ex-railwaymen show up during the week too, helping to strip down, clean and maintain the locomotives. 'You meet friends, see some of your old mates, and sometimes there are students here to pitch in with the work, which is good because it helps pass on the skills to the rising generation.'

I break off to go and see the extraordinary silver coffee pot fashioned in the shape of a locomotive, used in the very swish Swindon station refreshment rooms, which Brunel apparently always avoided. The *Illustrated London News* of 18 October 1845 praised Swindon railway station, notably its refreshment rooms, as 'perhaps second to none in the kingdom. Their accommodation is of the most elegant and splendid description.'

Time is ticking on and there are more ex-railwaymen to meet. Gordon Shurmer, born in 1921, was thrilled to receive a little clockwork engine for this fifth birthday from his father. It came in a little box, with a guarantee for sixty days. It is still working today, he says, and still kept in its original little box. He and I are back in the cab of *King George V*. While only a sixteen-year-old, Gordon was made a fireman at the Swindon works in 1939, and an engine driver in 1954. He drove both steam and diesel trains for more than twenty years. After a hip operation, he retired in 1982. Of the two, which type of power would he choose to work with, steam or diesel, I wonder? 'Steam every time,' he replies. 'That provided the biggest challenge, mentally and physically. With diesel, everything was on tap.'

Gordon remembers, during the Second World War, when working as a fireman on a train travelling near Dauntsey, a German plane flying so low it was almost running alongside them. 'The pilot had the audacity to wave at us. I was so scared I didn't need a laxative for a fortnight!' On another occasion he remembers watching the engine and tender of a moving train starting to part – they had to be hauled into Paddington by another engine.

Fred Simpson soon joins us. He and Gordon have known each other since 1937, both of them starting out at Swindon works as engine cleaners. They still work together now, as volunteers, at STEAM. 'Anyone can drive a steam engine – you could do it,' Fred says. 'There's nothing to it. Stopping it is where the skill comes in, listening to the brake box – using your sense of hearing, adjusting the valve on the vacuum.'

When Fred Jennings started working at Swindon, the signalmen looked after him like he was their son. He was fourteen, and the year was 1941. By the age of twenty, he had become the youngest guard anywhere in the Bristol division of the GWR. 'The great thing about working on the railway was that everyone was your friend and your mate. Having left for a few years to become the general manager of a company, what I found was that outside you were on your own. The railway had been the making of me. I was one of seven children. At home, I had to fight for everything. At work, I was treated as an individual. They taught you responsibility, integrity, and the need to always tell the truth. Trying to meet the standards they set has always served me well.'

Peter takes me to see the Wall of Names – more than 2,000 nameplates honouring people who worked on the GWR, many more than when I last saw it five years ago. Finally, he shows me a life-sized model of Brunel – based on the famous photograph of him against the massive chains of the *Great Eastern*, taken in 1857 by Robert Howlett – and the display case containing some of Brunel's personal paraphernalia: a walking stick-cum-gauge, his theodolite and tripod, drawing board and T-square, a wooden wheelbarrow and a silver bed warmer. And the cigar box that followed him everywhere.

What a fantastic day. Forget the locomotives, the real stars of the STEAM Museum are this band of ex-railwaymen who gather there every other Saturday to help tell the story of the GWR, and their own small roles in it, and pass it on to new generations. Beat a path to their door while they're all still hale and hearty is my strong advice.' While it was the merchants of Bristol who put up most of the capital for the GWR (mainly to help the city compete with Liverpool), and mobilised the movers and shakers of the age to get the necessary legislation through Parliament, Swindon was and remains its spiritual home. A visit to this museum, the railway village and the entire vast complex of the former works, witnessing both the continuing dereliction and the hotspots of sensitive regeneration, not only reminds us of Swindon's once world-class status. It also shows the town's potential to achieve greatness again perhaps, not simply by acquiring more shopping opportunities, corporate headquarters and bland housing estates.

Nationalisation of the railways after the Second World War marked the start of a period of steady decline, accelerated in the 1950s and 1960s as car ownership spread. Dr Richard Beeching, appointed chairman of the British Railways Board in 1961, published his famous inquiry report, *The Re-Shaping of British Railways*, two years later.

It revealed that one-third of the rail network was carrying only 1 per cent of the traffic, and that maintaining services in these areas could not be justified. About one-third of the country's 7,000 stations were closed under his report's recommendations. In 1986, BREL announced the closure of the Swindon works after 143 years of operation through two world wars. As a junior reporter for the Press Association in Parliament, I remember the shockwaves reverberating around the Palace of Westminster; and Labour MPs filling the Commons chamber to denounce Margaret Thatcher's government for its cruel betrayal of a loyal workforce, which would tear the heart out of a proud, hard-working town.

Nearly twenty years later, I learn that the closure had been on the cards for some time, and many employees had already left in search of more secure employment and better prospects.

* * *

Between the Bockworst Hot Dog stall and Lisa Marie's Number One Donuts stands Swindon centre's statue of Isambard Kingdom Brunel, and it is playing classical music. It was built to commemorate the inauguration of the first stage of the Brunel shopping centre and unveiled by Sir James Jones KCB, permanent secretary at the Department of the Environment, on 29 March 1973. 'Have you ever heard music coming out of the Brunel statue, Alison?' I ask.

'Never,' she replies.

A lady from one of the fast food stands comes over to stroke the dogs and she tells us that hearing the same music being played over and over can get a bit irritating.

We walk past the Brunel Rooms entertainment centre, go underneath the Brunel west car park and follow a section of restored Wilts and Berks Canal at Kingshill. The canal was built between 1795 and 1810. It was 52 miles long, starting at Semington near Melksham on the Kennet and Avon Canal, and went via Swindon to Abingdon on the Thames. An old canal milestone indicating 'Semington 26 miles' is about all that remains visible of the canal through Swindon's shopping centre.

The Wilts and Berks was used mainly to transport coal from the Somerset coalfields to London, along with locally produced bricks, clay pipes, building stone and agricultural goods. It prospered during the construction of the Great Western Railway, which, once built, grabbed much of the canal's business. In the late nineteenth century traffic dwindled and the Wilts and Berks became heavily silted. In 1997, the Wilts and Berks Community Group, now the Canal Trust, was formed with the ultimate goal of restoring the canal to a continuous navigable waterway. Between 1998 and 2002, the Trust joined a partnership with the Wiltshire Wildlife Trust, the Great Western Community Forest and Swindon Borough Council to restore the canal from Beavans Bridge to Kingshill. The project was funded substantially by a heritage lottery grant, and works included conservation of unique wetland habitat, creation of parkland, play area, excavation of the canal and reconstruction of Beavans Bridge.

They've done a grand job, I must say, and as part of the development of Swindon's Front Garden, another 2½ miles of canal are due to be restored. I suppose this will provide some consolation for the loss of so much of the rural character of the area.

We walk underneath the rebuilt Beavans Bridge to cross the old Midland and South Western Railway and the path we'd used to walk into Swindon. Alison recalls seeing the progress of the reconstruction of the bridge on her cycle rides into work at the old Princess Margaret Hospital. 'This is the first time I've seen it since it was finished,' she says. 'Magnificent.'

Once safely over the M4, we head west towards the landfill site just past the garden centre off Junction 16 of the motorway. We pick our way to the top of the hill south of the railway line between Swindon and Chippenham. The route of the public right of way isn't easy to make out around here, and a landfill site worker tries to put us right. Eventually we find the canal path into Wootton Bassett, another restored and re-watered section that makes for pleasant and easy walking. Old trees along the canalside have been cut down and the stumps burnt off. We reach the newly restored Chaddington Lock, very impressive-looking and a great opportunity for the dogs to enjoy a paddle and a drink. Further along we find a well-placed bench to rest up for a few minutes.

On the footbridge over the mainline railway, we look down at what remains of Wootton Bassett station, which was opened in 1841 and closed in 1965. The Beaufort Arms, away to our right, reminds us that this is also where the old Badminton line used to strike off towards Patchway, north of Bristol. The latter, otherwise known as the Bristol and South Wales Direct Railway, was built to provide a short cut for rail services between Swindon and the Severn Tunnel to avoid the more circuitous route on the main line via Chippenham, Bath and Bristol. Long-running campaigns to reopen the stations both at Wootton Bassett and at Corsham, between Chippenham and Bath, have inspired glimmers of hope every few years, following years of hard campaigning, only to end – so far at least – in disappointment.

'It wouldn't take much to reopen the station, would it?' notes Alison as we survey intact platforms, momentarily deafened by the roar of a high-speed First Great Western train flashing by underneath us. 'The road entrance is still there, and there's plenty of space for new buildings. Why don't they just do it?'

'Maybe one day.'

We go through the entrance to Lower Greenhill Farm, a set of imposing black iron gates flanked by two stone eagles, and soon pass Little Park. One of the establishment's dogs, a golden labrador, joins us and sees us off the premises several fields later when we reach the road into Tockenham. A Hercules plane from RAF Lyneham rumbles overhead, a deer leaps through a field of weeds, and the skies darken over Cherhill Down and the Lansdowne monument away to the south. We fetch up, all of us utterly drenched, outside the Londis shop in Lyneham, where Alison's husband Howard has come to pick us up, and we dive into the car.

* * *

From the bus stop near the main gates to the RAF base, the dog and I spring northwards towards Bradenstoke, past an enormous pile of gravel. Two women and a small girl are standing on top of the gravel mound watching planes taking off, their skirts blowing in the wind. Near the Bradenstoke Methodist Church, I pass two workmen painting some windows ready to install in a house, and listening to the midday news on their radio.

'The International Olympics Committee has the pleasure of announcing . . .' says a disembodied voice, '. . . that the Games of the 30th Olympiad 2012 are awarded to . . .'

'Paris,' said the fair-haired workman, confidently filling the pregnant pause, sitting back on his haunches.

'. . . the City of London,' says the IOC spokesman. The fair-haired workman shakes his head in disbelief.

'That's going to be hard for Paris – that's the third time in a row they've lost out,' he says glumly.

'Good for London, though, eh!' I say, winking at his younger workmate.

Moving on, I glimpse spectacular views between the houses on my right across the Avon valley towards the Cotswold Hills. The chap who runs the Bradenstoke village stores and post office is in the middle of doing his accounts when I call in to buy some fruit and a drink. Standing by the counter, I am loath to disturb him while he's totting up the figures, his back turned away from me. He is using the daylight from a side window to go through his paperwork. I eventually affect a cough. Having attracted his attention, I mention the story I'd heard about William Randolph Hearst and the vanishing priory down the road.

'The newspapers described it as an act of national vandalism,' he sighs, mournfully, Biro in hand. 'Come back another time and I'll show you some old photos. After the Dissolution, it was never used for religious purposes again, but even so . . .'

Bradenstoke Priory, founded by Augustinian monks in 1142, was dissolved in 1539. In the early 1930s, William Randolph Hearst, the American millionaire press baron, and the real-life Citizen Kane, provoked a public outcry by having much of the partially ruined complex of buildings dismantled and transported to St Donat's Castle, near Llantwit Major, South Wales. He had bought the castle in 1925 and planned to restore it using suitable materials, salvaged elsewhere. Questions were asked in Parliament about the legitimacy of the sale to Hearst, and how he was able to 'plunder' such an important piece of English heritage.

From the flagpole rising from the top of the ivy-covered priory tower, the Union Jack flutters in the breeze. Further along I can see that some of the medieval perimeter walling has recently been repaired and recapped. There are plans for a nature reserve around the medieval site.

Between the RAF Lyneham Saddle Club, stables and paddocks, and Crash Gate 6 there are more wonderful views through the trees into the valley. Cars and lorries on the M4 are tiny specks in the mid-distance, and trains glide across the wide valley like slow-worms. Near the corner of the airfield lie an abandoned plane, a real rust bucket, with its wings clipped, and the wreck of a fire tender with shrubs growing out of it. The distinct chill of postwar austerity appears to hang about the place, a semi-deserted Ballardesque landscape of runways, communication towers, the occasional lonely truck passing through. Lyneham is the home of the RAF's fifty Hercules C130K planes, but barring an about-turn by the Ministry of Defence, not for much longer. The base is due to close by 2012, the aircraft and their crews to relocate to RAF Brize Norton in Oxfordshire, with the overall loss of nearly 600 local jobs.

Seconds after getting airborne, the planes clear the patch of woodland near the end of the runway and lumber overhead in the direction of the Bristol Channel. They seem so close to the ground; you feel you could almost reach up and shake hands with the pilots, although it probably wouldn't be a good idea to try. The path on the edge of the hay meadow down to Foxham Lock is still muddy and slippery after the rain, but it's beginning to dry up in the warm breeze. A green woodpecker flaps between some hawthorn bushes.

The visibility is good today: familiar landmarks the Cherhill White Horse and the Earl of Lansdowne obelisk stand out clearly on the top of the downs near Oldbury Castle.

Near Foxham Lock there's a notice pinned to a gate about a new riding route running north-east alongside the canal. Stepping out on to Foxham Common, see clouds of butterflies and bees busying themselves around hedgerows laced with pink dog rose.

Heading towards East Tytherton, with the Foxham Inn behind us, we pass a herd of slumbering pale brown cattle, oblivious to the almost constant rumble of planes overhead. A deer scuttles out of the ditch by a hedgerow.

In East Tytherton, a Moravian church, housed in a huge red-brick building with a stone-tiled roof and tall red chimneys, makes an impressive sight. According to a sign outside, this was originally a preaching place of the eighteenth-century evangelist and hymn writer John Cennick. Mothers are picking up their children from Maud Heath's Primary School, so I guess it must be about 3.30 p.m.

A wealthy widow, Maud Heath of East Tytherton donated land and property for the building of a causeway in 1474, so that people could transport their animals and produce across the flood plain to Chippenham market 'dry shod'. It ran for 4½ miles from the top of Wick Hill to Chippenham Clift. A monument to her memory was created in 1698 beside the river bridge at Kellaways, roughly midway between Langley Burrell and East Tytherton. In 1838, the Marquess of Lansdowne and William Bowles financed a second monument on Wick Hill, the seated figure of Maud Heath, perched atop a high stone column, staring into the middle distance towards Kellaways. The inscription reads:

> Thou who dost pause on this aerial hight
> Where Maud Heath's Pathway winds in shade or light
> Christian Wayfarer in a world of strife
> Be still and ponder on the path of life.

In his Buildings of England series, Nikolaus Pevsner, another Wiltshire resident, who was buried only a few miles from here, comments rather acidly: 'The quality of the poetry matches that of the statue.'

The footpaths crossing the first couple of fields I need to traverse from East Tytherton towards Stanley are serviceable enough, but the next is a tangle of old rape crop and mare's tails, head high. Trying to thread my way through this lot is hard going, although the dog, being of short stature, encounters few problems apart from impatience at my slow progress.

Next, we find ourselves walking through the pasture of two large and very frisky horses, kicking their heels in the air, and cantering in our direction. Just in time, I manage to dive over the stile to exit the field, and Fly scrambles underneath the fence. 'They probably thought you'd come to feed them,' says a chap I meet who's driven up the side of the next field with the horses' afternoon feed.

'Well, I've got an apple in my rucksack. Perhaps they got wind of that,' I reply. Again the footpath ahead seems to have disappeared, so we skirt the edge of the field of newly planted rape, following some tyre tracks.

At Stanley I follow a minor lane and look out for the old branch railway line running between Calne and Chippenham, now part of the National Cycle Network Route 4. Having found it, I reach for the camera in its case attached to my belt to take a picture of the distinctive Sustrans signpost, indicating the mileages for Chippenham and Lacock to the north-east, Calne and Avebury to the south-east. But it's gone. My brand new digital camera, and all the photos I've taken since leaving Paddington station, has vanished and I'm absolutely gutted. I bet it's lying somewhere in that jungle of a rape field.

I stomp in weary resignation along the Sustrans route towards Chippenham, cursing aloud, furious at my negligence. Unused to hearing such language, the dog waggles her ears back and forth, trots along at my heels, sniffs at the vegetation, occasionally looking up at me in a solicitous fashion, as if to say: 'For heaven's sake, lighten up – it's just a camera.'

* * *

The next morning we retrace our steps from Bradenstoke in a kind of 'Groundhog Day' re-enactment of the previous day. At East Tytherton, I spot a notice pinned up outside Maud Heath Primary School announcing its proposed closure, owing to falling rolls. There are just thirty-five pupils at present, it says. Strange I should miss that yesterday, but then I see today's date at the top of the notice. The school appears empty already.

'Fly, find our camera,' I say to the dog for about the hundredth time. 'Where's camera – go find! Fetch camera!' The walking stick carved out of viburnam by Mum's neighbour George Allsop, a lovely man – retired farmer – comes in handy as I whack away at the old oilseed rape stalks that cover the footpath in that wretched field. Fly keeps disappearing and winding her way back to me through the forest of dead stems.

After about an hour of this apparently futile activity, the dog suddenly plonks herself in front of me and will not budge out of the way. Normally when Fly roots herself to the spot, it's because she's heard a strange noise and needs the kind of reassurance that only comes from being reconnected with me via her lead. Gently I tread down some of the stalks rising like bamboo canes around her, bend down and pull back the metal clasp at the end of her lead, to hook it round the D-ring of her collar. The dog promptly goes into reverse gear, disappearing through my legs, and nearly pulls me over.

Lying on a patch of damp soil in front of my boots, lies one black camera case, glistening with slug trails, containing a camera. Having found a bit of tissue paper in my pocket, I pick off the slugs, one by one, wipe off the morning dew on my trouser legs, and give the dog a pat. Good dog.

'Let's go, Fly. Find Chippenham. Go fetch. Let's go. Find the path. Good dog. Where's Chippenham?' Having ambled off and disturbed a deer, the dog returns to the task in hand. In the evening, I call in at my Mum's house in Lea, near my home in Malmesbury, to tell her about my loss, and the dog's recovery, of the camera. 'Have you heard about these bombings in London?' she asks, turning up the volume on her TV. 'The worst since the Blitz, they're saying.'

'What bombings?'

We watch the footage of searches and rescues around four locations in London. So far thirty-eight people are known to have been killed and hundreds injured, and scores of police have begun searching the wreckage for clues.

5

Box Hill & the

Corsham Quarries

Chippenham to Bristol

In which we find Mr Brunel's bed, locate the eastern entrance to the Box Tunnel by unexpected means, and are astonished by connections with the Second World War; the dog becomes unwell and is much missed. Historical aspects of the manufacture of chocolate and brass are noted.

In its original incarnation, Chippenham railway station was a modest single-storey affair with a pitched roof, designed purely to serve main-line traffic between Swindon and Bristol. However, as plans took shape for a Wilts, Somerset and Weymouth Railway, linking north Wiltshire with the south coast of England, it soon became clear that the station needed larger accommodation. A Chippenham-based engineer, Rowland Brotherhood, was contracted by the GWR to carry out the necessary developments. His firm had already built the foundations for the Wharncliffe viaduct, and other sections of the GWR between Twyford and Reading, and Bath and Bristol. Astutely, Brotherhood had followed the work westwards.

Now if you walk out of the station entrance, turn right down to the mini-roundabout and go right again, you'll see Brotherhood's home from 1842 on your left, immediately before the railway viaduct. Orwell House, built in the early nineteenth century, was used to receive Brunel on his visits to oversee and inspect the building work, particularly on the viaduct. It was a substantial property, featuring a croquet lawn, formal gardens, and later a west wing, to accommodate Brotherhood's extraordinarily large family. He and his wife Priscilla had fourteen children.

In 1864, having successfully led the fight for a united Italy, General Giuseppe Garibaldi visited England and was greeted by large cheering crowds wherever he went. Such was his heroic status, the general had already become one of the few military leaders, possibly the only one, to have a biscuit named after him. The Bermondsey-based manufacturers Peek Frean had first marketed their fruit-stuffed slabs of oblong-shaped, detachable biscuits three years earlier.

But I digress. On learning that General Garibaldi's train would be coming through Chippenham, pausing for a few minutes at the railway station, some of the many Brotherhood brothers decided to honour the general by firing a cannon volley in greeting. Such gestures were almost commonplace in Victorian days, apparently. Unfortunately, their aim was not 100 per cent accurate and the cannon-ball shattered the glass roof of the railway station.

For much of the 1990s, Orwell House was all but concealed from public view by the extremely tatty-looking frontage of the Dreams beds empire. Mercifully, this eyesore is in the process of being removed and the whole building is undergoing a programme of renovation and restoration. The network of grim pedestrian subways that sprawled underneath Brunel's viaduct was filled in recently, and the public space around it revitalised by new street-level footways, cycle paths and decent lighting.

Chippenham station has also undergone a £550,000 programme of improvements that make the place a pleasure to use. The internet café with its Box Tunnel fireplace, created by Rudloe Stone, and the restored Brunel office building next door are well worth a look.

* * *

A couple of miles west of Black Dog halt, the dog and I are happily rejoined by Alison Griffiths and her collie Badger, for the next section of our walk – to Corsham via Lacock. Unfortunately, I'm in charge of the route-finding today. It's another swelteringly hot day, and after a few wrong turnings (it must be the brain-addling effect of the heat), I find the bridleway that will take us up to Pewsham Locks. Derry Hill, thickly wooded, rises gently to the west, and we can see Golden Gate, the entrance to Bowood House, set in rolling acres of parkland designed by Capability Brown.

'Did you know that Dr Joseph Priestly discovered oxygen at Bowood House, Alison?' I ask.

'I do now,' she replies, affecting mild interest. 'How did he manage that then?'

'Well I don't know, but he was tutor to the sons of the 1st Marquess of Lansdowne, and he worked in the laboratory they had there. That's all I know.'

At Pewsham Locks some sections of the brickwork have recently been repaired and the length of this part of the Wilts and Berks canal path down to Reybridge, near Lacock has been opened up for walkers and cyclists. A man with a very long beard, sitting astride a quad bike and evidently a river warden, chugs past us. Alison and I briefly discuss how satisfying it might be to earn a living as a river warden, and what qualifications one might need to become one. It's a lovely morning, with plenty of drinking opportunities for the dogs along the brook that runs into the canal. Approaching Reybridge, we pass the well-fortified grounds of Ray Mill House and a policeman guarding the entrance, before crossing a field with horses towards the village. Camilla Parker-Bowles (now HRH the Duchess of Cornwall) moved to the area after her divorce in 1995.

We come into Lacock via the ford. The water level is low enough – a couple of inches – for us all to splash our way through it so that the dogs can cool their toes. We emerge into the village, now mostly owned by the National Trust, near the church and the pottery. A few visitors are doing what they do best in Lacock, which is pottering about. Lacock is the perfect place in which to potter about, gaze at the buildings and into people's gardens, whose surplus produce is often offered for sale to passers-by.

Having refreshed ourselves at the National Trust café, we walk back along Church Street so that I can reacquaint myself with a rather special fifteenth-century inn, named At the Sign of the Angel.

A week's research in and around Lacock for a magazine article first brought me here a few years ago. The wonderful oak panelling, sumptuous food and elegant furnishings made it a memorable visit, and ever since I've been on the alert for opportunities to return. At last, I have one.

After waiting a bit while a group negotiates complex arrangements for lunch for forty a few weeks ahead, I ask a woman staff member: 'Have you still got Brunel's bed by any chance?'

'We have indeed,' she replies.

'Great. I don't suppose I could have a quick look at it, could I?'

She goes to fetch the hotel's owner, George Hardy, who kindly agrees to show me the bed. He leads me up a narrow winding staircase, and asks a chambermaid: 'There's no one in this one? No – OK.'

'You're in luck,' he tells me, throwing open the door to reveal the great man's magnificent, lavishly carved mahogany bed, almost as big as the room itself, certainly the biggest bed I've ever seen.

George believes that he was aged about three when his parents bought it locally at an auction in the 1950s, following the dispersal of family belongings by Brunel's granddaughter Lady Celia Noble: 'Family history says they paid fifteen guineas for it,' says George. 'Apparently the bed was provided for Brunel's use when he was abroad doing some consultancy work on the continental railways.'

'In his spare time!' I interject.

'Yes, he seemed to be one of these people who managed to squeeze forty-eight hours' work into twenty-four. Anyway, he liked the bed so much that he had it shipped back to Britain to use at his home in London.'

Since it came to the Angel at Lacock, the bed – measuring 6 foot square – has been reupholstered. 'I've always understood that it was Spanish. But when my antique expert friends come here, there's always a huge debate about whether it is Spanish or French. Personally I think it is Spanish.'

George kindly agrees to sit on the bed to pose for a picture. 'I might get in trouble with the cleaners.'

'Well – don't tell them, then – or just blame me. Is it quite an attraction then, Brunel's bed?'

'Certainly, yes. People come here specifically to sleep in it,' replies George. 'There was a couple staying here just a week or two back. A lady had brought her husband here, a man in his sixties, a Brunel enthusiast. He was absolutely delighted to have slept in the bed belonging to his hero.'

As well he might be. It's a glorious room too, with a bathroom going off it, into which I step backwards gingerly to try to get the whole bed in the picture. It's not possible, unless I climb into the bath, which is probably pushing it a bit.

'Did you remember to ask where the old halt used to be?' asks Alison once I'd rejoined her and the dogs outside.

'No, but I've got some great photos of Brunel's bed and its owner.'

From Lacock, we head further west towards Corsham along footpaths. Having stopped, looked and listened, as instructed by the sign, we cross the Chippenham to Trowbridge railway line. At New Farm, we stumble across three dilapidated railway carriages, which appear to be used for storage, and a traction engine in a barn. Having entered a field full of curious horses near Thingley, of various shapes and sizes, we can't find a way to cross into the next. As the horses begin to encircle us, we hurl ourselves over some barbed wire. The path around the two sewage farms marked on the OS map seems to have done a disappearing act, too.

We pass many long-horned cattle; the horns are beige with black tips. The fields are quite small round here, and the air is pungent with the smell of freshly cut hay, which a few tractors are turning over ready for baling. Skirting the main line again, the yellow-flowered ragwort is rampant.

Alison: 'You know the story about ragwort?'

Me: 'Only that it's poisonous to horses. Do tell.'

Alison: 'Well, the ragwort originally came from the slopes of volcanoes, which are very dry and cindery. Someone bought a specimen back to the Oxford Botanical Gardens, and from there, the seed managed to get out, and because they thrive on these dry cindery conditions, exploded along all the railway lines of Britain.'

Me: 'Amazing. Look, I think we're coming into Corsham. Let's go find the old railway station.'

* * *

Scanning the track and an old goods shed from the footbridge, we look at the desolate spot for a few moments, before cutting down to the path that runs alongside the line in the direction of Box and Bath. Despite being built on the edge of only a modestly sized town, Corsham station bustled with commercial and passenger traffic for most of its years in operation. Initially, it served as a transfer point for passengers for Bath and beyond who were too scared to travel through the Box Tunnel, preferring to journey by coach and horses to rejoin the train at Box station. By 1864, an annual 100,000 tons of locally mined stone were hoisted on to trucks and transported from Corsham station. Even the weather sometimes pulled in large numbers of visitors. In December 1879, Corsham station swarmed with skaters, several hundred of them arriving each day: Corsham Park lake had frozen and Lord Methuen had agreed to allow people to skate on it.

It's worth remembering that the building of the GWR was commonly opposed, and treated with suspicion and contempt. Thousands of people lost income or were put out of business altogether with the coming of rail travel – coaching inns, toll bridge operators, stage-coach proprietors, turnpike trusts. While the decline of the canals was relatively gradual, the arrival of branch railways killed off some coaching companies virtually overnight, particularly when the Royal Mail began to switch its contracts from road to rail operators. In her excellent book *The Great Road to Bath*, Daphne Phillips records the winding-up of Corsham's turnpike trust in 1870 and the sale of its toll houses at Blue Vein, Pickwick and Lacock for £50 apiece; and points out that it was the smaller communities along the Great Bath Road who were often hardest hit.

The daily coaches, with their regular stops, stream of passengers and endless demand for horses had been the life-blood of the road. Without them it was desolate, its facilities largely disused, its innkeepers closing their doors, and their servants idle . . . In the late 1840s, huge sales were held of coach horses, coaches, post-chaises and pieces of harness. Many unwanted vehicles never found a buyer and were left to moulder and rot in deserted coach houses and inn-yards.

Vested interests aside, some ordinary travellers hated the railways too, and simply boycotted them. In a letter to a friend describing a journey from Wiltshire to London in January 1844, a lady called Marjorie St Aubyn wrote: 'My state of mind would not permit me to travel by rail for Mr Brunel is no hero of mine, since he has destroyed posting, put down coaches and compelled people to sit behind his puffing monsters.'

We soon find welcome shade from the mature trees alongside the railway line. The railway cutting west of Corsham, not as deep as the one at Sonning, is carpeted with yellow, pink and mauve wild flowers, swaying in the light breeze. We peer in vain through the thickening vegetation growing around and through the metal fence. 'According to the map, the tunnel entrance must be around here somewhere,' I say, pointing into a thicket, 'because this other footpath here is meant to go over the top of it. Hang on a minute, did you hear that?'

We don't so much see the eastern entrance to the Box Tunnel as hear it. Alison and I gravitate towards a spot on the path where the sound of trains rushing past us, just a few yards the other side of the thickets of bushes and trees, stops abruptly. Then – just birdsong. We walk over the top of the entrance to the tunnel to see if we can view it from the other side of the railway line, pressing our noses through the bars of the high metal security fence. Again no luck. Then a flurry of activity. Two railway workers clad in their yellow fluorescent jackets are descending steps on the other side of the line towards the tunnel mouth.

Nipping back to the other side of the line, I take my chance and slip through the unpadlocked security gate, tiptoe down the steps, and spend a few second gazing at the tunnel entrance. Not as ornate as the honey-coloured Bath portal, which I'd viewed several times in the past, the Corsham entrance is impressive nonetheless, quite austere-looking, faced with pale grey stone and dark slate-coloured bricks. Hearing voices below, I creep back up the steps. One of the railway men spared me a few minutes to clear up the yarn I'd heard about the number of tracks disappearing into the tunnel at the eastern end, exceeding the number emerging at the Bath end, almost 2 miles away.

'Now, it's the same number – two sets of rails going in, and two coming out,' he replies. 'But there was another tunnel entrance, a spur for goods traffic when the military moved into the area in the 1930s. That was used during the Cold War when a huge underground complex developed. There were lots of nuclear bunkers in the area, but the biggest of those was under Box Hill too.'

As we continue along the footpaths towards Westwells Farm and past the entrance to RAF Rudloe Manor, Alison and I discuss when exactly the Cold War ended.

'Fall of the Berlin Wall?' Alison suggests.

'Must have been – late 1980s?' I reply. 'I remember Ronald Reagan banging on about the Russian Bear and the Four Horsemen of the Apocalypse in the early 1980s, so it must have been after that.'

We ponder on what it might be like being cooped up in a nuclear bunker for months on end. 'Just imagine never having any natural light coming in, and think of the smells down there after a few days with all those other people.' I venture. 'I'd rather be dead. A great white flash, mushroom cloud, and that'd be it. Dead.'

We look across the landscape of uncultivated fields and scrub that runs along the top of the Box Tunnel. A green woodpecker flashes past us. 'So how did they make this tunnel then?' Alison asks. 'Did they dig into it from the top – or along it from the sides?'

'Well, both really, as I understand it. Several shafts were driven into the limestone along the top of the hill, and at the bottom of these, they would then start digging, and also from the cuttings at each end of the hill.'

It took five years (1836–41), some 1,200 navvies and 400 horses to build the 3,312yd-long tunnel for the Great Western. The men got through a ton apiece of gunpowder and candles a week to blast a way through the stone, often working in gruelling, ghastly weather, sometimes right through the night. More than 100 navvies died, and many more were injured, in accidents during the construction period. It would appear that Brunel did not permit such casualties to lie heavily on his conscience. Once, he was shown a list of 131 navvies admitted to Bath hospital with serious injuries sustained while deployed on building the Great Western over an eighteen-month period between 1839 and 1841. Brunel's comment was characteristically sanguine: 'I think this is a very small list considering the very heavy works and the immense amount of powder used . . . I am afraid it does not show the whole extent of accidents incurred in that district.'

Passing the guarded entrance to RAF Rudloe Manor, Alison spots what looks to her like the top of one of the airshafts into the Box Tunnel – a large grey mushroom-shaped structure, 8–10ft high, perhaps. I ask one of the guards on the gate into the base, and he confirms what Alison suspects. Having reached the western edge of the plateau across the top of Box Hill, we flop down on the GWR bench overlooking the valley as it stretches away to Bath and soak up the panoramic views. We decide to catch a bus later from Box to the information centre at Corsham to find out more about the area.

Despite the early hour, about 11 a.m., Mark the barman at the Quarrymans' Arms on Box Hill makes us some excellent cappuccinos. He tells us how the building of the Box Tunnel, and the advent of the GWR, led to a rapid expansion of the local quarrying industry in the late nineteenth century. This, in turn, enabled the military to colonise what the quarrying industry, following its heyday, had left behind – vast areas of empty space, linked by a well-developed network of tunnels. A huge subterranean military village began to take shape, largely in secret, during the mid-twentieth century. There were some 80 miles of tunnels under Box Hill alone, he says.

We cross over the A4 and drop down into the By Brook valley, the rooftops and church tower of Colerne on the north-western horizon framed by a cloudless blue sky. The shade along the footpath, between the meadows and the By Brook, towards Box Village, provides relief.

From the footpath, the biggest swan either of us has ever seen waddles unsteadily across the grass in the grounds of the converted mill buildings that house Peter Gabriel's Real World recording studios. At the end of the excellent tree-lined dog walk at the foot of the recreation ground, we pause to admire the water garden and wildlife area donated by the Lovar Foundation in 1997. Unfortunately, it being Tuesday, the library housed in Selwyn Hall is closed. So I call in to one of the village stores asking the whereabouts of the house where the Revd Wilbert Awdry, author of the Thomas the Tank Engine series grew up. Keith, the shopkeeper, explains the location, near the Box Tunnel entrance, adding that it is now a guest house called Lorne House: 'So he had the view of the railway and of the tunnel from where he lived. Did you know also that Samuel Taylor Coleridge lived in the village for a while?' He points me in the direction of the Box Parish Council office, a few doors along, for more information. 'Just tell Margaret that Keith sent you.' Margaret Carey, clerk to Box Parish Council, patiently listens to me wittering on about Brunel's bed that I'd seen at Lacock, and what George Hardy had told me about his parents acquiring it at a local auction. He'd mentioned Box and Brunel's granddaughter Lady Celia Noble in this context. Could she shed any further light on this, perhaps? Margaret took my details and said she would make some enquiries and get back to me.

Wilbert Awdry (1911–97) spent most of his childhood in Box. It was supposedly his boyhood memories of hearing the railway engines 'talking to each other' along the GWR around Box that inspired his famous stories.

Alison and I go and look at Lorne House on our way to the Box Tunnel viewing point off the A4, a short distance up the hill. We contemplate the tunnel's Bath portal, beautifully framed by the steep rise of pasture, slumbering cattle and woodland fringing the top edge of the hill. The plaque says: 'Box Tunnel – Engineer Isambard Kingdom Brunel. Constructed 1836–1841. Length 3,212 yards. West portal cleaned and restored in 1986 by British Rail Western Region, assisted by the Railway Heritage Trust, Wiltshire County Council, Box Parish Council and private subscriptions to commemorate its 150th anniversary.' From the viewing point, part of the tunnel entrance is concealed by vegetation growing on the cutting, but I can see it perfectly from the middle of the bridge that carries the A4 over the railway line.

At Corsham's Visitor Information Centre, we study a display of extraordinary historical descriptions and old photographs telling the story of the MoD's once secret underground world. Some 2 million tons of limestone rubble were dug up and removed between 1935 and 1940 to create a vast subterranean military base. Up to 15 tons of ammunition were handled every day, and 100,000 tons of it were stored underground near here. There was also an underground railway station, an aircraft factory – at nearby Spring Quarry – and an internal telephone network linking all the underground installations connected to the GPO at Clift Quarry on the A4; about 30,000 sq. ft of office space was created underground with lift access to the surface.

Underneath RAF Rudloe Manor on the A4 (now Rudloe Hall Hotel), from the former Browns Quarry, No. 10 sector control group conducted its activities in the Second World War. There was a plotting room for charting positions of the various parts of the armed forces. By 1943, the central ammunition depot at Corsham encompassed some 125 acres of subterranean chambers, containing 300,000 tons of explosives and munitions. The total cost of the depot was more than £4.5 million – somewhat over the £100,000 originally allocated for the 6 acres envisaged at the outset. A standard-gauge railway branched off the GWR main line at the Corsham portal of the Box Tunnel to service the various ammunition storage districts and the rest of the underground village. There was a platform, refuge sidings, and a narrow-gauge railway system with diesel locomotives, turntables, engine houses and workshops.

On another display, under the heading of 'Myths and Legends', we read that in more recent times, the military underground sites are said to have been used for the secret storage of captured UFOs and a strategic reserve of more than 100 steam locomotives if ever there was a national emergency requiring locomotive power using local coal instead of imported fuel. Gob-smacked and boggle-eyed, we step out into the fierce afternoon sun, the dogs in tow, and go in search of a cup of tea, a mission which we are unable to accomplish.

* * *

Alison has introduced me to her friend Sue Alexander, another keen walker, who is accompanying me, Fly and Badger on our way from Box to Bath. It's another lovely, sunny morning, a little cooler than of late. As we emerge from the ladies' loos opposite the Blind House, Sue asks an elderly passer-by why it's so called. The Blind House, that is. 'That's where they used to lock up the wrong uns,' he says. 'Called a blind house 'cos it got no windows.' We pass the library and get on to the Macmillan Way following the northern banks of the By Brook and the railway line towards Shockerwick and Batheaston. We pass the very grand-looking Shockerwick House, now a residential care home.

On the outskirts of Batheaston, Sue points out a large clump of a pink-flowering shrub in someone's front garden. 'Himalayan balsam,' she says. 'It's one of these alien species that's spread everywhere, and is killing off some of our native plants.' We cut through a modern housing estate to avoid walking on main roads, and get some clear directions from a pair of elderly residents for getting under the A4 and the main railway line, and on to the canal path. Once through a field of newly shorn sheep, we go over the branch line running down to Bradford-on-Avon, and soon emerge on to the Kennet and Avon canal path. A stationary heron monitors the activity from the other side of the water – a gentle ebb and flow of walkers, boaters, joggers and cyclists.

The cool of the morning has given way to a sweltering hot afternoon, and the canalside garden of the George pub on the eastern edge of Bathampton lures us in for a tray of tea in a shady spot. The dogs share a bowl of water under the table and flop on to their sides to doze. At Bathwick, we come off the canal path and descend into Sydney Gardens, where there are many beautiful bridges over the railway. On a day like this there is only one way to appreciate the view of these bridges, and the trains going back and forth – from a comfortable bench, ice cream in hand.

* * *

Margaret Carey, clerk to Box Parish Council, has been back in touch with me with some more details about Brunel's granddaughter, Lady Celia Noble. Lady Celia (1870–1962) lived in Bath for many years at 22 Royal Crescent. She was born to I.K. Brunel's daughter Florence and Florence's husband Arthur James, an assistant master at Eton College. Following Celia's marriage to Saxton Noble, who later succeeded to a baronetcy, she became a well-known society hostess, arranging concerts of chamber music, and receiving visitors at the Crescent including Princess Marie-Louise, a granddaughter of Queen Victoria, and Queen Mary, consort of George V, during the Second World War. She died in 1962 at her home in Bath, aged ninety-two.

* * *

It's 8 a.m. and a staff member at Bath railway station travel and information centre is unlocking the door and beckoning me in. 'Morning. I wonder if you can direct me to the Avon Walkway from here,' I ask. He frowns.

'The Avon Walkway?'

'Yes, I'm walking to Bristol – starting off on the Avon Walkway, along the canal.' 'You're walking?' he asks, incredulous at the very notion. 'No idea,' he concludes, raising his hands in surrender. Overhearing us, the man behind me in the queue tells me to turn right out of here, right again under the railway line round back of station and I can't miss it.

The dog's gone to Alison's today as she's under vet's orders to take it easy for a few days. She's been prescribed a fortnight's supply of antibiotics, each pill having to be concealed in a bite-sized lump of meat. Having company for so much of the walk so far has spoiled me. The prospect of a long day's walking on my tod, just me and my map is a little daunting. There's a good breeze, though, and the temperature comfortably warm.

The canal path heading west out of Bath is quite narrow and weedy in sections. Seeing the elegant arches of the viaduct at the foot of Lyncombe Hill, framed by clumps of trees in full leaf, marching up the hill, lifts the spirits. Over the road bridge, an information board describes the route of the Bristol and Bath Railway Path, which I scan briefly. 'From Bristol the route rises gently to Fishponds, running close to the built-up area while retaining a green character . . . views across open spaces and valleys . . . the impressive clay-bottomed viaduct . . . Mangotsfield station. Warmley station . . . gardens and grotto.' It interconnects at several points with the Avon Walkway, so I shall probably alternate between the two.

After the wharf buildings of red brick and stone, converted into flats, the sounds of a busker playing drift overhead, as I go underneath the footbridge to Sainsbury's at Green Park. Near the big gasholder and the Homebase complex over to the left I reach a signpost: Bristol 15½ miles.

Passing the *Bath Chronicle* building conjures memories of attending an interview, in 1980 or so, for a job as a district reporter in Radstock. My interviewer went out of his way to emphasise that Radstock was a quiet little place, where not much tended to happen, and that an ambitious young lass like me might find it a tad dull there. Having soon afterwards landed a job on the *Gloucestershire Echo* in Cheltenham, just in time for the arrest of an alleged Soviet spy in our midst, his advice proved timely. Cheltenham in the early 1980s was about the most sensationally exciting place a young reporter could wish to be, thanks to one Geoffrey Prime, the Russians' mole at GCHQ for fourteen years who was earning a living as a cab driver at the time of his arrest. He was sent down for thirty-eight years.

Huge clumps of wild buddleia in flower line the route for miles. There's a charming little packhorse bridge near the Dolphin Inn on the way to Weston Lock. Over to the left is Twerton, a small parish in the early 1880s that was partially demolished to make way for the GWR and then rebuilt. I remember reading a few years back, quite possibly in the *Bath Chronicle*, that Twerton was about to get its own poverty tsar. It became fashionable in the 1990s for government ministers to appoint 'tsars' to try and sort out various intractable problems – poverty, drug misuse, cancer. This had the benefit of deflecting attention from their own political failures to provide durable remedies on to someone else's shoulders.

A group of five young men on a narrow boat called *Sundowner* and I pass each other at regular intervals all the way to Keynsham. Back in the world of forbidding signs, I soon note that new degrees of privacy are being claimed. Bath Marina and Caravan Park has put one up on the footpath saying 'Strictly Private. No fishing. No Swimming.' For the next few minutes, I try to decide whether the use of the word 'strictly' is in any way meaningful.

By the time I reach New Bridge, built in 1734 to replace a ford, I am well and truly back in open countryside. I pause to read the information board. The construction of New Bridge was part of a programme of works to make the River Avon navigable, and link two roads from Bristol to Bath. Until the bridge was built, Old Bridge or Churchill Bridge was the only one to cross the Avon at Bath. A trip to the Globe Inn from Bath by tram was a popular day out, apparently. The trams ran across New Bridge between 1904 and 1939. Despite the presence of railways and roads, the landscape here has apparently changed little over 200 years. The landowners and landscape gardeners of the late eighteenth century carefully sculpted and planted the grounds of nearby mansions so as to blot out the views of old coalmines in the area. When laying out the gardens at Newton Park and Kelston Park, Capability Brown is said to have been inspired by the romantic appeal of landscape paintings of the time.

During construction work on the Great Western Railway a Roman villa was uncovered in 1837 near this spot at Newton St Loe. Evidence suggests it was a large third-century farm, sited close to the River Avon for easy access to the thriving town of Aqua Sulis (Bath) and the Bristol Channel. Among the mosaics revealed was a circular one – the earliest of its kind found in Britain – showing Orpheus charming wild animals. The mosaic was reconstructed in 2000. It would appear then

The author and Fly at the start of their journey at Paddington station in front of the statue of Brunel. The statue, which was designed by the sculptor John Doubleday, was commissioned by the Bristol and West Building Society and presented to British Rail Western Region in 1982.

Anne Boston, who accompanied the author on the first leg of her journey, and a statue on the Grand Union Canal path near Paddington station.

The Brunel family memorial at Kensal Green Cemetery, London.

Bob Harrington (left) and Colin Sadler in the railway carriage café at the Wallingford terminus of the Cholsey and Wallingford Railway.

Susan Harrington tending plants in her recycled pig troughs at the Cholsey and Wallingford Railway.

The ticket office, Didcot Railway Centre, Oxfordshire.

No. 5051 Earl of Bathurst *at Didcot Railway Centre.*

Fire buckets, Didcot Railway Centre.

John Minchin, a volunteer at the Didcot Railway Centre.

Fly in a poppy field between Didcot power station and the village of Steventon, Oxfordshire.

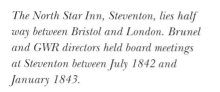

The North Star Inn, Steventon, lies half way between Bristol and London. Brunel and GWR directors held board meetings at Steventon between July 1842 and January 1843.

The author, Fly and Badger at the Brunel statue, serenaded by classical music, in the Brunel Shopping Centre, Swindon. The statue was unveiled on 29 March 1973 by Sir James Jones, permanent secretary at the Department of the Environment, to mark the inauguration of the first stage of the shopping centre.

A mannequin of a GWR typist, STEAM Museum, Swindon.

A mannequin of a female GWR riveter sitting inside a locomotive boiler, STEAM Museum.

Peter Pragnell, an ex-carpenter at the Swindon works, and his 'double' at STEAM Museum.

Friends reunited: Fred Simpson (left) and Gordon Shurmer at STEAM Museum. Fred and Gordon have worked together on the GWR, off and on, since 1937.

Alison Griffiths and Fly's best friend Badger at New Farm, near Corsham, Wiltshire.

Inset: A milestone near Box.

The glorious Box Tunnel, Bath portal, as seen from the A4 bridge at Box, Wiltshire.

that the building of the railway helped turn up interesting archaeological finds as well as extend understanding of geology.

Next to New Bridge there's a weather-boarded New England-style pub called The Boat House, deserted it being mid-morning. The path continues on the south side of the river, alongside fields of old unharvested crops of rapeseed, very brown and beaten down by wind and rain. These crops give off the rather pungent odour of piles of damp unwashed clothes. Hearing again a familiar blend of background sounds – birdsong and rumbling trains – puts a spring in my step, but it's not the same, walking without the dog. At the old railway bridge I climb up the embankment to join the Bristol and Bath railway path to take advantage of the shade.

At Saltford, I learn that in 1727 the Avon Navigation opened up the tidal river between the Bristol Channel and Bath. Before the installation of locks, the tide swept right upwards to Bath. This made the water salty at Saltford, unlike the water at Freshford – upstream from Bath – where it was fresh. Before the locks were built, goods were mostly transported by horseback as boats travelled so slowly. The newly navigable status of the river was not universally welcomed, however.

The livelihoods of local miners were threatened as cheaper coal could now be brought in from Wales and the Midlands via the Severn and Avon. Mill owners feared they would lose the flow of water that had powered their operations. In protest, a group of wreckers almost destroyed Saltford Lock in 1738, just three years after an Act of Parliament made the destruction of locks and weirs punishable by death. From about 1750, Saltford became a source of new wealth in and around Bristol as a prominent manufacturing centre for brassware.

This stretch of river here, known as the Shallows as you could wade across it to the other side during tidal times, is also famous for its eels. These slippery fish begin their life in the Sargasso Sea, east of the Caribbean, about 4in long with a glassy appearance. They grow during their long swim to Europe. Huge numbers are drawn to the Severn estuary where elvers move into rivers in their second spring. By the time they are about ten years old, and about 3½ft long, they return to the ocean and swim back to their breeding grounds to repeat the cycle and finally die. Eel traps (called ullies or hullies) were used to catch the fish for selling on for the cooking pot, helping the locals to earn a living. There are picnic benches all along the river here – a very green and pleasant spot.

Back on the cycle path, I amble towards a narrow bridge overhead, about a mile or so north from Saltford. Underneath the bridge I see a couple peering at a sign, and drawing closer, layers of rock exposed on each side of the old railway path. Apparently, these rocks represented the floors of the ancient seas that covered England 200 million years ago during the lower Jurassic period. The fossil evidence shows that the sea would have been warm and relatively shallow, supporting large fish-eating reptiles. The sea floor was probably soft and muddy at the time, but gradually turned into layers of limestone, clay and shale. Members of the Bristol Naturalists' Society and the Bath Conservation Volunteers cleared the site to reveal these layers of stone.

On each side, thistles, cow parsley, nettles and teasels bob about in the breeze. Suddenly, apparently out of nowhere, there appears a set of newish-looking painted buffers, a length of single-track railway and a deserted platform with a sign saying: 'Avon Riverside'. There are well-tended plants in pots on the platform, it all looks spick and span. In fact, I haven't witnessed this level of spick and span-ness on a railway platform since inspecting Sue Harrington's planted pig troughs at the Hithercroft Road Industrial Estate in Wallingford. A working railway – worked by whom? I later find out why it's not marked on my OS map. It's a new railway – or rather a new station terminus for a restored section of the former Mangotsfield to Bath Green Park branch line. This line once formed part of the Midland Railway, and has been restored by the Avon Valley Railway after years of being told it couldn't be done.

I can see Bitton Church over to the right. Going over the river again, I miss the turn off to the Avon Valley Country Park, near Keynsham, but take a hairpin bend down to the river off the cycle path. Rejoining the Avon Walkway, I can see from this side of the meandering river something of what's going on at the country park just as well as if I'd taken the correct turning, possibly better. There are kids playing on see-saws, and tyre swings suspended from tree branches among the goats and Highland cattle. Very flat away to the right. Hay is neatly baled in the field. Path becoming very overgrown, hard to navigate. Feeling tired. Approaching Keynsham, there's a proliferation of moored canal boats, including one called *Festina Lente* – Move More Slowly. In the afternoon heat, who could do otherwise?

The first flock of Canada geese I've seen since skirting the edge of Coate Water bobs about on the Avon. Dozens if not hundreds of cruisers and canal boats seem to be moored here. Coming off the path to go over the road bridge and sink an apple juice at the Lock Keeper pub, just on the edge of Keynsham, there's a sign: Bristol 8½ miles, Bath 9½ miles. Have I really only got halfway through today's walk? I wish the dog were here to give me a tow.

The afternoon sun saps my energy. On reflection I'm glad the dog's not with me to suffer the heat of the day. Back on the River Avon Trail, I pass a community woodland, called Sydenham Mead, of broad-leaved trees, oak ash, field maple, small-leaf lime and alder. Near here, the Duke of Monmouth was tried for treason by the hanging judge, Judge Jeffreys, and beheaded. Monmouth's supporters were flogged, imprisoned or transported to Barbados. You see – this is what information boards do for distressed walkers; make you count your blessings, when you're suffering from imminent heatstroke.

The brass industry thrived in this whole area because of the nearby presence of calamine (zinc ore) from the Mendip Hills, and the copper smelted locally. Coal to drive the furnaces was mined in Kingswood. The water mills harnessed the power of the river to process corn, cloth, paper, lime, ochre and brass. The most important of these was the Avon Mill, now a pub. By the early nineteenth century this mill had become the centre of Bristol's thriving brass industry.

Now here's another information board – albeit one with a factual error – that resonates. 'Fry's relocated their chocolate factory from central Bristol to Keynsham in the 1920s, and to this day the company continues the eighteenth-century tradition of chocolate manu-facturing in the Bristol region.' Well, up to a point.

Aware of its commercial potential, an apothecary named Dr Joseph Fry began experimenting with chocolate and its derivatives in 1748 in his Small Street, Bristol, premises. As a result of his curiosity, J.S. Fry & Sons was born. Several years after the Quaker-owned firm merged with Cadbury's of Birmingham in 1919, manufacture switched to Keynsham. The workers were asked what they would like as the name for the new factory. Somerdale was chosen – the huge red-brick complex of buildings, with tall red chimney, that I see ahead of me. From the train, the giant Cadbury's sign can be seen today on the southern side of the site.

The Fry family had many connections in Wiltshire, many of its members originating in the county. How do I know all this? Well, when my mother fell and broke her hip a few years back she was fortunate enough to spend several weeks convalescing in Malmesbury Hospital. Her neighbour for the duration was an amazing lady, in her nineties, called Diana Fry, a member of the chocolate-making dynasty, whose son Johnny would regularly visit from his home in Scotland, laden with bars of Cadbury's chocolate.

My walking route loops almost all the way round the Cadbury's factory, and the car park, the sun glistening on their roofs, in a U-shaped bend. A woman sits smoking a cigarette on a step by one of the factory exits, her feet crossed at the ankle.

Keynsham has an Augustinian Abbey, founded by William, Earl of Gloucester in 1166. It was dissolved in 1539, the monks surrendering their possessions to Henry VIII. The Abbey stone was used to construct many of Keynsham's later buildings.

I'm on the Monarch's Way now as well as the Avon Walkway, a little sign tells me. A goods train hauling about twenty wagons is on its way to Bristol. I see the tower of Keynsham church, and thickly wooded banks away to the right. Such is the profusion of butterflies, dragonflies and smaller airborne creatures buzzing about my face, I have to swat them away.

Having passed underneath a huge electricity pylon, Hanham Lock soon hoves into view, the first of 106 locks on the Kennet and Avon Canal. Two neighbouring pubs, The Chequers and The Old Lock and

Weir Ale House, are doing a brisk trade with lunchers seeking shade from the sun. The ground is beginning to rise quite steeply on either side of the river here – so this must be the start of the Avon Gorge I guess. The railway line is elevated here, hugging the edge of the gorge. Giant hogweed and wild geranium rampage everywhere. The lads on the *Sundowner* narrow boat pass me again, and we wave at one another like old friends.

Avon Valley Woodlands has an abundance of mainly young trees, mixed with a few fragments of ancient woodland between Conham and Hanham. Horse chestnuts, ash, hawthorn, sycamore, together with areas of scrub and open land to encourage wild flowers. Around a century ago it would have all looked quite different. Each side of the gorge was extensively quarried, the rock faces exposed. Spoil heaps and waste tips scarred the hillsides. The last quarry closed in the 1930s. At the Conham River Park end of the gorge, extensive replanting began in the 1970s. Elsewhere, the regeneration of the woods has been entirely natural, with some sycamores growing out of old slag heaps. The brass industry dumped much of its waste along the banks of the Avon, seeping into the river and causing major pollution. When, in 1749, the Bristol City Corporation ordered it to stop the practice, the industry started to recycle the slag by reheating and moulding it into useful building blocks.

I come across a little ferryboat service over the river to Beeses – a tea gardens and pub, operating on Fridays and weekends in the summer months. There's a bell to ring if you want someone to come and fetch you. Even though it is neither a Friday nor a weekend day, I'm sorely tempted to clang the bell, partly out of devilment but also – being a social animal – I've not had a decent conversation all day to distract me from worrying about the dog.

The river has worn a large groove through the Pennant sandstone, heightening the dramatic scenery of Hanham gorge around me. Conham Hall, demolished in 1971, and surrounding woodlands were used as a refuge for persecuted Baptists and other Nonconformist Christians in the eighteenth century. Copper ore was imported from Cornwall by sea, and then brought up the Avon to Conham. At this time the river would still have been tidal at Conham, with barges travelling here only at high tide. Conham was also on the edge of the Kingswood coalfields and there were pits at Hanham and Crusehole. The copper works flourished. There were thirty furnaces operating here by the 1720s, and the zinc works was also open. Zinc and copper were the two metals needed to make brass, and brass mills flourished all the way along the Avon to Bath.

* * *

There's some attractive new housing, three storeys of red brick, on my right at St Anne's Park. Soon a light industrial landscape, somewhat littered and unkempt-looking, takes over from the green woods of the gorge, although the buddleia continues to run riot. Once, the area was known for heavier industry – steelworks, cotton and tar plants – and the Netham Chemical Company alone spread itself over 40 acres. There were ample supplies of locally mined coal. Netham Lock was the gateway to the navigable Avon and the canal network running east from Bristol's floating harbour.

Until the nineteenth century, Barton Hill was a rural parish, but in the 1950s and '60s it was selected for the construction of several tower blocks of flats; the earliest of these has been replaced by new terraced housing. Over a road bridge and looking for a footpath underneath the railway line, I walk past a depot where dozens of brand new traffic cones and road signs are piled high on wooden pallets, awaiting despatch. No entry. Road narrows. No parking. Men at work. Slow down. It's a driving test candidate's dream.

Underneath the railway line, trains roaring above, large dollops of water drip on to my head. I pass a dumped washing machine, and from here the Avon becomes visibly tidal, with clear tide marks on either side – green vegetation above the line, washed-up brownish silt below. The colour and consistency of the water remind me of the gravy my mother used to make for the Sunday roast, the silty banks glinting brightly in the sunlight. Looking at the black-headed gulls and fulmars bobbing about on the surface of the water or swooping up and down and around it, the river landscape's swift transition to marine environment appears almost complete.

The path is heavily overgrown and I'm completely covered in nettle stings from head to foot. Someone ought to get down here with some loppers, because the Avon Walkway is meant to be a proper pedestrian route. By 3 p.m. my legs feel like lead, as I trudge dutifully through more trading estates into the centre of town. Passing the Bristol Dogs' Home, I glance through the bars of the blue metal fence into one of the compounds, and immediately wish I hadn't. Five pairs of hopeful canine eyes stare up at me and one pair of front paws is plonked high against the fence separating me from the dogs. This is quite unbearable. I have to avert my gaze and hurry on. One dog is quite enough to look after. I wonder how she is?

6

Capital Works

Bristol to Bridgwater

In which the heatwave damages the rails around Bristol, the SS Great
Britain *is prepared for reopening, Tyntesfield and Clevedon bring much
delight; the dog's attire is suitably upgraded at Burnham and we face a
quandary at Hewish.*

No fan of the new railways, Augustus Welby Northmore Pugin,
the Victorian aesthete and architect, derided Brunel's Bristol
Temple Meads as a 'mock castellated work' with 'all sorts of
unaccountable breaks' in the station buildings. Though a pale shadow
of Brunel's original three-storey extravagance, the current station
entrance retains a castellated, turreted façade and provides an
uplifting vision of Gothic grandeur. Opposite the entrance to the
station, in a bland modern block of buildings, is The Reckless
Engineer pub, featuring a silhouette of Brunel wearing his stovepipe
hat, etched into the window. From his earliest experience of the city,
Bristol provided both palette and canvas for some of Brunel's finest
masterpieces of design and engineering – ships, bridges, docks,
railway terminus – and was transformed by them.

This morning rail services have been disrupted after a section of
track had to be closed. Some of the rails had bent in the heatwave,
and the Up and Down trains to Paddington are having to share a
single track. From Temple Meads station, I loop round to Cattle
Market Road to join the Avon Walkway, taking me under the
railway, past the Gardiner Brunel Rooms and the Brunel Garden
Centre in Straight Street. While walking through Castle Park, the
first of many temptations and distractions hoves into view: the
tethered GWR FM Big Balloon from which passengers can float
500ft above the city. Further along the hillside is Sallyport, a

thirteenth-century hidden exit from the Castle to a ditch outside. The main purpose of Sallyport was to defend Barbican Gate. The bones of someone's pet monkey were once unearthed here – evidently an Indian species.

Over Bristol Bridge down Welsh Back, past the floating restaurants, Glassboat, Spyglass, Il Bordello. Redcliffe Quay over the other side of the river. Lovely plane trees rise above the cobbled walkway, offering welcome shade. In Queen's Square, criss-crossed by good wide paths, sunbathers have already yielded to the mid-morning heat. Some of us pedestrians stand in a little group to watch Prince Street Bridge being swung open to let through a motor cruiser. Further west along the floating harbour, Brunel's Buttery provides refreshments to the visitors.

Never a Roman town, Bristol's history as an important port dates back to the eleventh century. Bristol Bridge and Docks were built in 1287, trading in salt, wool, fish, wine, sherry and rum. By the late eighteenth century, the ships sailing in and out of Bristol were heavily involved in the 'triangular trades' exporting guns and metalware to Africa, transporting slaves across the Atlantic, returning with more slaves, tobacco, rum and molasses. In 1809, the floating harbour increased granary trade and trows brought Welsh stone. Shipbuilding began in 1837 with Brunel's SS *Great Western,* and later the largest iron-hulled ship of its time, the SS *Great Britain.* Further west along the harbour, I pass some modern apartments painted in pastel shades. A great deal of construction is going on over the river – half-built blocks wreathed in scaffolding and plastic sheeting.

At the Maritime Heritage Centre preparations are nearing completion for the grand reopening of the SS *Great Britain* on Saturday. A mounted policewoman is patrolling the area. The restored ship, resplendent in colourful rigging, is breathtaking. I'm told at the reception desk that guided tours finish tomorrow – after that it will be self-guided audio tours only. Fortunately, I was able to walk the decks of the SS *Great Britain* and view the interiors, with the benefit of a real tour guide, Bob Evans, just a few weeks ago when the last licks of paint were being applied.

Brunel's famous ship, launched in 1843 in the presence of Prince Albert, was the first screw-driven, steam-powered, iron-hulled vessel the world had seen or, as the *United Service Gazette* of that year put it, the greatest sea-going vessel 'since the days of Noah'. After more than a century of noble service, carrying thousands of passengers to the USA and Australia, and troops to fight the Crimean War, the ship was left to rot for decades in the Falkland Islands, but finally towed home to her original dock in Bristol in 1970.

An £11 million lottery grant has funded a complete refurbishment, and a sophisticated dehumidifying system, underneath a plate-glass 'sea', helps to preserve the iron hull. The bold and imaginative restoration seems to complement the ambition of the original design perfectly. What I appreciated most was the way in which the story of the ship's chequered history is told largely through personal accounts of the passengers and crew, images of them and their families experiencing life on board. Four million people have visited the ship since she returned to the very dockyard where she was built. She's also proved a very popular venue for wedding receptions and anniversary parties, not least among the descendants of people who sailed to New York in the glory days of her first voyages across the Atlantic.

My tour of the SS *Great Britain* took place after a conference I'd attended at the Bristol Watershed Centre. Organised by Brunel 200, the project has not only been planning events and celebrations across Bristol and the South West to mark the bi-centenary of the engineer's birth, but also helping find and nurture the polymaths of the twenty-first century. The place overflowed with contemporary stewards and interpreters of Brunel's legacy.

Tristram Hunt, broadcaster and author of *Building Jerusalem: The Rise and Fall of The Victorian City*, kicked off the proceedings with such a barnstorming performance, full of passion and insight, that I almost forgot that it was Sir Isaac Newton that he had championed in the BBC Great Britons poll back in 2002. He spoke of the self-sufficiency, respect for architecture and civic vision that sustained many of the Victorian cities like Birmingham, Manchester and Liverpool, 'not defying London but ignoring it'. He urged planners to be brutal with developers who would cover the countryside around our cities with concrete sprawl.

Next up was the novelist Mavis Cheek, whose eponymous hero in *Patrick Parker's Progress* has an obsessional devotion to Brunel. Before giving a reading from her book, she described the latter as 'a bit of a bastard' for the way he dumped his first love Ellen Hulme; and sparked a lively exchange with some of the civil engineers in the auditorium on the question of women bridge builders and why there appeared to be so few, if any, of them.

Matthew Tanner, director of the SS Great Britain Trust, told us that as part of the restoration of the vessel, the Trust had tracked down a company that could manufacture and supply a genuine 'vomit smell' to help give the sick bay added authenticity. Professor Colin Taylor, of the Department of Civil Engineering, University of Bristol, pointed out that Brunel and his colleagues had to design and create cranes and other tools to get their projects built, and that there was something of Alan Sugar's management style about the engineer. The scale and nature of Brunel's innovative thinking were comparable with those of the Space Race in the 1960s.

We heard from Leslie Perrin, chairman of Brunel 200, whose law firm Osborne Clarke had acted as solicitors to the Great Western Company on the land purchases. He had just arrived from Paddington by rail and the next morning would be flying direct from Bristol International airport to New York, thus helping to realise Brunel's dream for a permanent integrated transport system between London, Bristol and New York. Raising the profile and status of engineering, and attracting more women into the profession, were identified as prerequisites for removing today's barriers to creative technological innovation. As the author Francis Spufford put it so well, we need to find better ways of encouraging people who, like Brunel, 'appoint themselves to re-imagine the world'.

Andrew Kelly, director of Brunel 200, had no doubt that were he alive today Brunel would still be working to advance and harness technological knowledge in the realms of health and the environment. 'Great conference – so interesting!' I remarked to the chap sitting next to me after we'd filed back into the hall from the coffee break. He crossed his legs, folded his hands together and stared up at the ceiling.

'Hmmm, yes,' he replied, arching an eyebrow and adding a second or two later, 'Angus Buchanan isn't here, you know.'

'Well, perhaps he's got something else on,' I replied, flicking through the conference papers balanced on my lap.

'He's regarded as having written the *authoritative* biography of Brunel,' my neighbour continued, pressing his fingertips together, as if in prayer.

'Well, I thought Adrian Vaughan's biography was a *terrific* read.'

'Hmmm – very rah-rah Brunel . . .'

There was something else he thought I ought to know. 'You do realise that the Clifton Suspension Bridge wasn't actually built to Brunel's original design? Substantially modified.'

'It was certainly built in tribute to him though, wasn't it – I always think of it as Brunel's bridge.' The man fell silent, our conversation clearly serving no further purpose.

* * *

My route via the Avon Walkway continues south from the SS *Great Britain*, along Gasferry Road, across the floating harbour and joins the path running alongside the New Cut. Although I'd visited the Clifton Bridge in the past, I'm about to get my first view of it from the water's edge. Having crossed the river just short of Brunel Way, I head north-west along the path and suddenly there it is ahead of me, spanning the deep gorge, a perfect marriage of form and function. A few cars glide back and forth, as if by magic. Closer up, I can see a workman clinging to the side of the bridge, and after a few minutes, dangling from it.

A chap bowls up alongside me, dismounts from his bike and looks up. The tide is low this morning, the water silty and brown. 'Amazing, isn't it?' he says. 'The scale and the height of it.'

'Quite incredible,' I reply.

At the age of twenty-two, and resident engineer to his father Sir Marc's Thames Tunnel project, Brunel was seriously injured in a flooding accident. Six men drowned, and he was struck by a falling timber, sustaining internal injuries and a badly damaged knee.

While recuperating among the fine Georgian terraces of Clifton in Bristol he heard of plans for a competition to design a bridge across the Avon gorge. Brunel wrote that he had been 'struck by the grandeur of the idea and of the splendid effect which might be produced by a bridge of this construction, were it to harmonise with scenery so peculiarly suited to such a work of art.' In March 1831, one of Brunel's designs, based on the suspension principle, was finally selected by the committee overseeing the project, despite the initial verdict of the judge, Thomas Telford, that it was so flawed in its conception as to be impossible. For years the project was dogged by financial problems and rioting on the streets of Bristol over the Reform Bill. Work on the bridge soon lost its momentum and was only completed in 1864, five years after Brunel drove himself into an early grave.

* * *

There's a bit of a mud bath just past the suspension bridge at the junction of the walkway with Nightingale Path, which I join to walk uphill into Leigh Woods. Halfway up I meet a man from the local water company who asks me if I've seen any evidence of a burst pipe.

'Someone phoned in with a report of water pouring down the hill.'

'Well, there's a lot of mud at the bottom of the path, but I've not noticed any pouring water. Do take care if you're going down there – it's quite steep and slippery.'

'As if I haven't got enough to do,' he mutters.

At the top of the hill, there are a few vehicles parked along a dark lane. One is a van that has written on the side of it: 'UPS – Synchronising the World of Commerce', and whose driver is sitting at the wheel eating sandwiches and reading a newspaper.

The 80-acre Leigh Woods were given to the National Trust by the tobacco baron Sir George Wills in 1909, both to protect them from housing development and to secure public access. The northern parts, which I've been walking through, contain some ancient woodland and rare native trees such as the Bristol whitebeam, although over at Paradise Bottom stands a 108ft high giant redwood. Stokeleigh Camp Iron Age hill fort, constructed in about 300 BC, also lies in this northern part of the wood. Once past Abbots Leigh, and a wonderful Gothic-style property called The Priory, my first view of the sea, north-west beyond Avonmouth towards Wales, puts a spring in my step. Soon cars and lorries on the elevated section of the M5 over Avonmouth appear as little coloured specks in the distance.

In Fifty Acre Wood, numerous paths straggle off in all directions and at some point I think I must have walked on most of them trying to find the way out. There must be worse ways of spending an hour than going round in circles in Fifty Acre Wood but none springs to mind at present. Wandering through unfamiliar landscape, with not another soul about to ask for directions, sometimes has its drawbacks. Having inexplicably stumbled across the Corus Hotel off Beggar Bush Lane, I regain my bearings, with the help of the reception desk staff, and soon I've found the Monarch's Way. This path traces the flight of Charles II after the Battle of Worcester in 1651. Pursued by Oliver Cromwell's army, the fleeing monarch travelled first north then south through the Cotswolds and the Mendips to the south coast, and then along the South Downs to Shoreham before escaping to France. The trail links existing paths and bridleways for 600-odd miles.

Back in open countryside, the heavens open as I look down towards Long Ashton on my left, and pass a field of snoozing white pigs. Skirting the southern flanks of Ashton Hill Plantation I emerge on to a narrow lane, which takes me down to a busy road running past the North Somerset Showground. I walk up Belmont Hill, past Belmont Farm into some woods. There's a bit of a path tracking the edge of the woods, but it's quite overgrown, which makes for very slow going. On the other hand it's preferable to walking on that busy road in pouring rain and risking life and limb, especially since I can hardly see through my spectacles.

The rain subsides and I give the specs a polish. A fox mooches about ahead of me, sniffing clumps of wild garlic, his red coat and big bushy tail flecked with black.

After passing a large house called Belmont, swathed in scaffolding, there's a good clear path ahead and soon I am back in civilisation in the form of small groups of people ambling about in leisurewear carrying guidebooks and cameras. Cars are parked nose to tail along the track, and a couple of coaches. Through the trees I can just make out parts of a large red stone building and, much to my relief, realise that I've arrived at Tyntesfield.

Studying the front of this magnificent building, it's quite affecting to recall how, within a matter of weeks, the National Trust was able to raise the £20 million needed to save Tyntesfield for the nation, following the death in 2001 of the 2nd Lord Wraxall, George Richard Lawley Gibbs. It was described as the last major High Victorian house and estate to have survived, largely unaltered, with much of its original furnishings intact. From the start, the fund-raising campaign captured the public's imagination, and the donations rolled in – more than 77,000 individual contributions were made. A local bus driver raised £1,500 from passengers on his route, and donors included American servicemen hospitalised at Tyntesfield during the Second World War.

In the shop at the back of the house, a National Trust volunteer says they're still only able to offer pre-booked tours at the moment, and gives me a leaflet with details for applying for a place. Fair enough. Given my soggy and bedraggled appearance, even I wouldn't allow me into Tyntesfield. I buy a guidebook to read in the café.

The house that stands today was rebuilt in 1865 and funded from the considerable wealth amassed by William Gibbs and his family from the mining and shipping of guano – solidified bird droppings – from the islands off Peru. The fertilising properties of guano made it a valuable agricultural commodity. In 1843, two years after the opening of the Great Western Railway, William Gibbs made several visits to the original house at Tyntesfield, built in 1813, while he was staying with his sister Harriett down the road at Belmont. He bought

Tyntesfield from the widow of its builder later that year. It is believed that the proximity of the house to the GWR may have been a deciding factor.

There was a strong family connection with the railway. William's brother George Henry Gibbs was a director of the GWR, and a staunch ally of Brunel in the latter's many battles. When worried shareholders were calling for Brunel to be replaced as the company's engineer, Gibbs stood by him. Unfortunately Henry died in 1842 while on a visit to Venice, leaving a son Henry Hucks who went on to become the Governor of the Bank of England and the 1st Baron Aldenham. He also left a diary detailing the day-to-day challenges and triumphs during the railway's planning and construction.

* * *

I wander round the back of the house and step into the chapel. Suddenly, being unable to view the house today seems unimportant, as I can't imagine it's more stunning than the chapel. A devout High Church Christian, William Gibbs built several churches, including two dedicated to St Michael and All Angels, at Paddington and Exeter, as well as the chapel at Keble College, Oxford.

On my way out to pick up the path towards Nailsea, I spend a few minutes exploring the orangery, kitchen gardens and walled garden. The orangery (1898) has been closed off, covered in scaffolding and put on the Buildings at Risk Register by English Heritage. As a Listed Building, it has to be protected until funds are available for complete restoration. The flowerbeds are going well – in particular the sweet peas, winding their way up individual bamboo canes. Chrysanthemums of different colours stand to attention. I'll have to come back and see the rest another time.

Once over the Bristol to Nailsea road, I join the footpath to Watercress Farm, a sad collection of abandoned buildings and signs alerting passers-by to their dangerous state. I'm walking along a buried trackway, much overgrown, getting filthy – eventually emerging on to Backwell Bow.

A man standing by the road is looking through binoculars. Says he's looking out for pigeons to shoot. Are they a pest then? Yes, they eat the corn and the barley. So do you eat them then? Yes, pigeon casserole. What's that like? Lovely – tastes a bit like beef.

* * *

From Nailsea, the path running alongside and north of the railway line soon veers north-west and diagonally across fields, then due north to join a bridleway. Once on the bridleway, heading west, large areas of bracken (the first I've seen so far in any quantity) and open land appear, and dog walkers exchanging morning greetings and gossip. The paths are well defined and signposted, as they have been almost all the way from Paddington. My dog is still on sick leave, and once again I am missing her company.

The North Somerset Levels and moors, a 40-square-mile area of open low-lying landscape and wetlands, drained by long straight canals or rhynes, is completely unfamiliar territory to me, and all the more alluring for it. As well as draining the land, these man-made rhynes also serve as 'wet fences' separating hundreds of square or irregularly shaped fields, creating a patchwork of miniature islands. Many of these rhynes are rich in wildlife, particularly rare insects.

As a result of sympathetic farming and land management, otters are making a comeback to the area and a North Somerset Otter Group has been formed to monitor their progress and numbers in the coming years. The waterlogged nature of the levels also makes them archaeologically significant as the conditions enable the preservation of organic remains such as leather and wood.

After walking along Nailsea Wall, I break off to the right, crossing Blind Yeo (a river), to find the footpath running north towards Tickenham Road and the motorway, past Clevedon Moor. I realise I may have gone wrong only at the end of a field from which there is no easy exit, just a heavily overgrown gate. It would have been sensible to retrace my steps back to Blind Yeo and check the map more carefully. Instead, I spend fifteen minutes or so beating back nettles and brambles sufficiently to allow me to hoist myself over the gate. A lady out for a morning horse-ride sets me straight and shows me an alternative route up to the road, to avoid having to traipse back to the river and start again. Within a few minutes of passing underneath the M5, I'm standing at the entrance to the medieval manor of Clevedon Court. Having bought my timed ticket, and awaiting the moment when I can present it, I settle down on a bench in the terraced garden to read the National Trust guidebook. Ten generations of Elton baronets and their families lived here until the property was accepted by the Treasury in part payment of death duties, and passed to the Trust in 1960.

Built in the thirteenth century by the Norman de Clevedon family, the manor was bought by the 1st Earl of Bristol in 1630. Confiscated in the Civil War, and handed over to the son of Sir Walter Ralegh, the property was later recovered by the earls of Bristol, only to be sold on again. In 1709, a fabulously wealthy Bristol merchant Abraham Elton, who made his money from brass making, became the new owner of Clevedon Court and he was later made a baronet. But it was the philanthropic ideals and benefactions of the Victorian Sir Arthur Hallam Elton, the 7th baronet, JP and MP, that made the town of Clevedon what it is today – a modest but elegant resort of distinctive character, with an exceptionally fine pier. Sir Arthur spearheaded campaigns to provide the town with a piped water supply, public lighting and sewers, as well as schools, churches and a cottage hospital. He was also a friend of William Gibbs at Tyntesfield.

A family of prodigious and diverse talents, each generation of Clevedon Eltons managed to distinguish themselves one way or another. Classics scholar and poet Sir Charles Abraham Elton, 6th baronet, was disowned by the family after he married Sara Smith, daughter of a Nonconformist Bristol merchant. They produced thirteen children, including two sets of twins.

Dame Dorothy Elton, wife of the 9th baronet, amassed a large collection of nineteenth-century glassware, notably painted rolling pins and cucumber straighteners. Then there was Sir Edmund, 8th baronet, the self-taught potter whose unusual styling and glazing of pots became internationally recognised and branded as 'Eltonware'.

There's a wonderful photograph of him in the guidebook, perched on a stool next to one of his works in progress, resplendent in deer-stalker hat, handle-bar moustache, white overalls wrapped around his tweeds and tied at the waist, paintbrush in hand and looking rather grumpily at the camera.

The first five baronets were all called Abraham, although the first to make Clevedon Court his permanent home was the Revd Sir Abraham V (1755–1842). He also had the distinction of being president of the committee formed to realise a long-standing proposal to build the road bridge over the Avon gorge between Clifton and Leigh Woods, and organise a competition for the best design – and we know who won that.

What most of the Clevedon Eltons appeared to share over the centuries was a tendency to spend the wealth created largely by the 1st baronet, and a disinclination or inability to replenish it. Inevitably, this marked lack of interest in keeping the family coffers topped up sealed their fate. Gradually, the dynasty became doomed to be dispossessed of its finest asset, the Court itself, their family seat. On such lack of financial acumen and entrepreneurial flair among the English landed gentry, and against a background of steadily declining incomes from farming and land holdings, the National Trust was founded in 1895, to save places of historic interest or natural beauty for the nation. As a result, members of the lower orders, like me, can visit such places and have a jolly good snoop round.

David Fogden has been the National Trust's administrator at Clevedon Court for the last six seasons, and there's not much about its history, and the Elton family's ancestry, that he doesn't know. 'They never invested in the railways, for example,' he tells me, as we survey the sprawling stone façade of the building from a spot near the south porch. 'And there were much grander families than the Eltons living in the area. Mostly, they were just too interested and busy with other things to invest their inherited wealth so as to secure the Court for future generations of Eltons. The 3rd baronet went bankrupt, one is said to have pursued illicit affairs with a number of ladies in Bristol, and others cultivated their interest in the arts and literature.'

Unlike some National Trust properties I've been in, where visitors are corralled and route marched with military precision from one room to another (and in one memorable case required to wear blue plastic bags over our shoes to avoid contact with the priceless carpets), Clevedon Court permits the public to wander at will, at their own pace.

After an introductory spiel from the lady volunteer looking after the entrance, most of the other visitors in my group disappear into the Great Hall whose walls are festooned with paintings of Elton family members. I amble off in the opposite direction, and stumble into the Old Great Hall, now housing the museum. The first thing I clap eyes on is the nameplate of steam locomotive 2937 *Clevedon Court*, displayed high on the stone wall above the hearth. There's a photograph of the engine on the next wall. Built in 1910 as one of the last of the 'Saint' class of locomotives, it was apparently acquired in the 1960s by Sir Arthur Hallam Elton II, 10th baronet, after being taken out of service along the GWR. In contrast to his forebears, Sir Arthur Hallam II actually earned his own living, as a filmmaker, and was also a fanatical collector of art and historical documents relating to the Industrial Revolution. The story goes that the retired locomotive *Clevedon Court* was transported by road from the Swindon engineering works to its namesake, but made it only as far as the Court's entrance gates. The steep gradient of the driveway, coupled with the sheer size and weight of the locomotive, combined to foil the attempted rehoming of the engine, and it had to be returned to Swindon, whereupon it was eventually dismantled. However, David Fogden readily confesses that the story has been called into question: 'We did have a visitor here not so long ago who challenged this version of events. He believed that the locomotive was actually broken up in 1953, so perhaps there's a bit of a mystery there that needs clearing up.'

Certainly, Sir Arthur Hallam II, who was at school with John Betjeman, was one of the more progressive, socially responsible and far-sighted of the Eltons. Although his father Sir Ambrose led a frugal existence on an estate that was by then mortgaged to the hilt, he'd done little in the way of forward planning either to secure the fabric of the Court or to shore up the family's dwindling resources. It fell to his son, Sir Arthur, to act to save the building before it was too late, and it was thanks to his efforts that the Society for the Protection of Ancient Buildings made a substantial grant for essential repairs.

When you climb the Queen Anne staircase at Clevedon Court, you'd be well advised to take your time, for here on the walls are mounted a tiny portion of the 10th baronet's collection of eighteenth- and nineteenth-century prints and lithographs of bridges, railways and canals. Exquisite depictions of the Clifton Suspension Bridge, Sir Marc Brunel's Thames Tunnel, Ivybridge, Bathford, Bristol Temple Meads may otherwise pass you by, en route to the bedrooms. The rest of the collection went to the Ironbridge Gorge Museum in Shropshire, in lieu of death duties again.

Talking of bedrooms, it took me some time to work out why the beds and chaises longues at Clevedon Court have tiny holly leaves placed bang in the middle of the covers – to deter visitors from plonking themselves on the furniture. It must come as a great relief for the current generations of Eltons, who still have homes at Clevedon Court (Julia Elton, and her brother Sir Charles and his young family), to enjoy some privacy there once the visitors have gone.

Outside, manning the refreshments stall, a friendly chap sporting a red beard, a straw boater and a Union Jack waistcoat, is trying to rustle up business for the cream teas, 'the best in the West' he claims to the dithering onlookers.

* * *

I head west into Clevedon past the hospital, the ambulance and fire stations on my right; the funeral parlour, the Salvation Army (1904) and the Village Hall Institute (1868), now Clevedon Town Council, on my left. The railway station, long gone, has been replaced by an area of small shops, a toilet block and a supermarket. Nearby is the Great Western Road, a charity shop called Brunelcare, and a genuine Curzon cinema, apparently one of the oldest in continuous use in the country.

My requests for directions to the tourist information centre are met with blank gazes. Finally someone points me towards the public library. At the library, my attention is directed towards some leaflets stacked against a wall. Having grabbed a handful, I learn that Samuel Taylor Coleridge and J.R.R. Tolkein spent their honeymoons in the town; and that famous former Clevedon schoolchildren have included Edith Cavell, the First World War nurse, and Sid Vicious, the eminent punk rocker and Sex Pistol.

The sea is sepia colour, quite rough today, but both Steep Holm and Flat Holm, small islands off the coast of Weston-super-Mare, are visible from the promenade. At the foot of Alexandra Road, near the entrance to the pier, is a beautiful Victorian drinking fountain, tiled in various shades of green. There's a plaque on it saying it was erected in 1895 by Mr T. Sheldon, restored by the Clevedon Civic Society in 1992 and is now a Grade II Listed structure.

Lacking both the infrastructure to accommodate large crowds of visitors, and the kind of beach that might otherwise attract them, Clevedon remains largely unspoilt, rich in Victorian and Georgian architecture – perfect pottering territory. The bandstand, built in 1887 to mark Queen Victoria's Golden Jubilee, is still in use, and the more modern buildings along the sea front, including the amusement arcade, are small scale and low level. National flags are strung out along the promenade. Further south, near the miniature railway, is the Marine Lake, which owing to its declining use over the years has been dubbed locally 'Stinky Corner'. There are some children throwing stones into it, and a few dogs taking a swim. A poster promises a re-enactment of the Battle of Trafalgar on the lake in the autumn; that should be worth seeing.

The dominant feature of the sea front of course is Clevedon's pier, an improbable architectural gem, certainly the finest I've ever seen, and a real credit to the community effort that sustains it. A peerless pier. What makes it improbable is its location, bestriding a sea of powerful currents and perilous tides, and its chequered history, which began, as so many things appear to have, with the railway revolution.

The close proximity of the new main line, and the building of a branch line linking Clevedon with Yatton in 1847, opened up the possibility of a through connection from London and Bristol to South Wales via paddle steamers. Having located and secured large quantities of redundant Barlow rail from Brunel's broad-gauge railway in South Wales, the contractors were able to start construction of the pier in 1867. By bolting the rails together, they were able to create the slim, elegant arches that supported the deck and gave the structure its simple uncluttered outline.

Thousands turned out for its opening ceremony two years later, and over the decades, enjoyed steamer trips to and from Barry and Penarth, firmly establishing Clevedon as a popular seaside resort. The collapse of two spans in 1970 placed the long-term viability of the pier in doubt until English Heritage and the National Heritage Memorial Fund made grants totalling £1 million to plug the fund-raising gap in 1984. During that year, Woodspring District Council granted a ninety-nine-year lease to the Clevedon Pier Trust, the green light for the complete dismantling and rebuilding of the structure.

Although the adjoining Royal Pier Hotel is now sadly boarded up, the pier today has long been restored to its former glory, after nearly twenty years of dereliction, a popular haunt for anglers as well as promenaders and boat trippers. You can take tea in the pagoda perched on the end of the pier, and look round the art gallery in the Toll House. Last but not least, unlike some brasher piers further down the coast, Clevedon's allows dogs. I love it.

* * *

Having wound my way around the headland, via the Poet's Walk, past the twelfth-century St Andrew's Church, I'm looking out for the best way for the dog (now restored to robust health, I'm glad to report) and me to cross Blind Yeo. Eventually, on a well-used stretch of recreational land at the back of a modern housing estate, I spot the footbridge, and once over the river, pass a caravan park and several 'Keep Out' signs.

Using a fallen branch and a confident manner to fend off a herd of curious cattle, Fly and I make our way over farmland south towards Kingston Seymour. Depending on the wind direction, occasional sounds from the motorway, and beyond it the railway traffic, help to keep us more or less on track.

Both the map, and a farmer I speak to, indicate a crossing underneath the motorway near Phipp's Bridge, but although I can see a little tunnel through which the rhyne passes, there's no sign of a path next to it or anywhere in the vicinity. Another farmer driving cattle in for milking suggests the footway does exist but may have been covered in water. So off we go to find the footbridge over the M5, a bit out of our way but the only option. We seem to zigzag about all over the place negotiating this area around the M5 and the railway before finally hitting Hewish.

To my alarm, I realise that not only has it taken us about three hours to walk as many miles, but my intended onward walking route of twists and turns, skirting this, that and the other rhyne, in a landscape almost devoid of human settlement, may not be wholly advisable.

I stop at the Quart Inn on the A370 to consider the options. The barmaid is a little hazy as to buses from here, as are her other customers in the public bar. The lady at the plant centre up the road is similarly uncertain, and no one seems to know if there is a pavement along the A road into Weston-super-Mare from here. Short of bright ideas, I start walking out of the village on the main road hoping that inspiration will soon strike. In the playground of St Anne's School four children are helping a blonde woman to put away some equipment and I throw myself and the dog on their mercy.

She and I spend about ten minutes in conversation about the best way of either getting back to Clevedon or going on to Weston. I could wait for a Bristol bus, then try and get out at Yatton and walk from there (4 miles). Or I could try and get a Weston-super-Mare bus to Sainsbury's at Worle, and then try and see if there was a direct bus from there. Sensibly Fly has decided to lie down with all this chat going on, enjoying the curious attention of the children. After a bit more consideration, one of the girls, aged about eight, then whispers something into the woman's ear. 'Look,' she says finally. 'I can drive you to Sainsbury's at Worle. There's a big bus turn-around area there.'

The six of us pile into her people carrier, the dog straddling my knees, and they kindly drop us off at Sainsbury's at Worle. While relieved to be no longer stranded at Hewish, I am not sure that standing outside Sainsbury's at Worle is a whole lot better. Eventually I make an executive decision: 'Sod the buses – we need a taxi.' A couple emerging from the supermarket with a packed trolley take me in hand, and give me a taxi number which I dial on the mobile.

On hearing the story of my poor route planning and fears of being forever marooned in Hewish, the taxi driver gives Fly a stroke, settles her on the floor behind the driver's seat and tells me about the habits of his one-year-old shih-tzu terrier. The schools are just about to break up for the summer. I expect the summer influx of visitors to Weston is about to hit the town?

'It's not like it used to be,' he says. 'Fifteen hotels have shut down in Weston in the last five years or so. People don't often come here for a week or a fortnight any more. It's mainly day trips or short breaks of two or three days.'

'Why's that then?'

'Well, the council's been dragging its feet deciding what it's going to do about the Tropicana and the pier.'

'People still like old-fashioned British seaside holidays, though, if the facilities are good and well maintained?'

'Yes – but it's also the sea at Weston, it's silted up, makes the water brown. There's so much competition now, with foreign holidays being so cheap. People can fly places for a few pounds now, can't they?'

* * *

It's a dull overcast morning, with a fresh sea breeze, perfect for walking. To find such relief from the summer's baking heat has been a rare treat for the dog and me so far. It's 9.30 a.m. and Weston's army of litter pickers are already out trawling the beach, dragging black plastic sacks behind them. The donkeys have been unloaded from a lorry and are being offered their morning hay from a round metal feeder driven into the sand. Some of the animals are having a good roll on their backs, legs waving in the air, enjoying their freedom before the tack goes on for their working day.

In the first week of July during the 1930s, Swindon became a ghost town. The place practically emptied as railway workers and their families headed for the seaside. Special trains were laid on to carry holidaymakers to the coast, places like Weymouth, Blackpool, Cornwall and Weston-super-Mare. Newsagents in such resorts stocked the *Swindon Evening Advertiser* for the duration, although there must have been precious little local news to report.

Miles of pale yellow sand stretch out ahead, as I look at the outlines of Steep Holm and Flat Holm, and Brean Down. The 'Seaside Award' part of the beach (in other words most of it) is closed to dogs between 1 May and 30 September, I read on a notice. So dog and I stroll along on the promenade, passing shelters and benches along the front and the deserted Tropicana, a walled open-air swimming pool which has been shut and empty for some time. A car park attendant is telling an elderly couple of the plans to redevelop and reopen it. Indeed, the current issue of *North Somerset Life* magazine describes the promised 'multi-unit entertainment centre providing a variety of attractions in a secure family environment'. These include swimming pools, five-screen cinema, health and fitness centre, hotel, skate park, cafés and shops. On the dog beach, beneath the sand dunes, a few walkers throw sticks and balls for their grateful hounds.

Soon I turn towards Uphill, via Links Road, and exchange greetings with a few golfers. A small field next to the road is grazed by alpacas and miniature horses; two men are repairing a stone wall. One levers off the old coping with a crowbar, while the other resites the stones.

The first sea defences at Uphill, which lies below the level of the highest tides, were built in medieval times. Since 1606, or perhaps earlier, a sluice or similar structure has protected the community at high tides, while allowing the river to flow out to the sea at low tide. Second World War prisoners rebuilt the sluice, and after the devastating floods of December 1981, it had to be reconstructed. In

2003–4, with mechanical parts wearing out again, the Environment Agency oversaw the creation of a new structure that met modern safety requirements. The gates close one hour either side of high tide during spring tides. We pass Uphill Boat Centre then skirt the edge of Uphill Hill, a Local Nature Reserve and Site of Special Scientific Interest, covered in flower-rich grassland that supports many insects, particularly butterflies. The limestone exposed here on the quarry face was formed 300 million years ago when warm shallow seas covered most of Britain. Shells of marine animals settled on the sea floor to create rocks known as carboniferous limestone. Since then huge faults in the earth's crust have lifted and folded the rocks to form the Mendip, the upland area spreading westwards to Brean and east to Cheddar.

Limestone was quarried at Uphill from the early nineteenth century to the 1940s, for building roads and to make railway ballast; it was also burned as quicklime. In 1826, quarrying revealed three caves that would have been used by Neolithic people during the Ice Age. Stone tools have been found in layers on the cave floors, together with the bones of animals eaten by these people, and of animals that occasionally ate them – including woolly mammoth and cave lion. The caves have since been quarried away. On top of the hill stands St Nicholas Church, dating from the eleventh century, without a roof since the 1860s.

The salt marsh stretching away towards Brean Down contains a number of nationally scarce plants, including sea barley, slender hare's ear and sea clover, creating splashes of colour in late summer; though a bit early to see much of this at present here, the hill will soon be yellow with buttercups and yellow rattle, wild thyme and bee orchids. What tantalises and teases any walker picking a path through this huge landscape is that so much of it is inaccessible. Straying from the paths is almost certain to spell trouble – the massive mudflats, sandbanks, river, sea and the endless rhynes will either send you scuttling back to safety, or swallow you whole.

We're having to take particular care between the River Axe and Middle Rhyne, walking a large semicircle round the back of the new

Weston-super-Mare sewage treatment works, pretty well the only building of any size between this point and the sea. I pause to look back at the fine views towards Uphill and St Nicholas Church. Flocks of Canada geese patrol the sewage works, and clouds of butterflies flutter among the teasels, thistles and cow parsley that bow to one another across the path, almost concealing it.

On the road, I go past Lympsham Forge, which makes fire escapes and galvanised caravan steps along with the more usual rose arches. While fighting my way through some brambles and nettles on the way to Berrow, I manage to lose the dog's lead. It was all my fault – I had to dive into a field next to the one with the footpath as the inquisitive cattle wouldn't leave us alone, then clamber in and out of a rhyne to get us back on track. So I pinch a length of baler twine I find wrapped around a gate to tie to the dog's collar.

Having found a lane that appears to be heading in the right direction, we pass the Animal Farm Country Park – 'Rain or Sun We're No. 1 for Fun' – and at a safe distance over to the right, the Big Wheel and some stomach-churning rides at the Brean Leisure Park. At a junction of roads on a green at Berrow, there are bus shelters with landscape murals painted on their sides. So much better than boring advertising (and no bus timetable information), which seems to be the norm on most urban bus shelters.

On to the sand dunes. Owing to the winds sweeping in off the sea, Berrow's fore-dunes comprise an ever-shifting, changing landscape, supporting a variety of coastal habitats for flora and fauna. Lyme and marram grass help trap the blowing sands. These and other plants grow rapidly when covered with fresh sand and form deep branching roots that bind with it. In these more stable conditions, other plants flourish. Nourished by rabbit droppings and decaying plant material, soil develops – emerging species include evening primrose in the summer, and sea buckthorn and yellow rattle. The latter is fairly easy to identify, and I'm wondering if the slightly sweet, honey-like scent wafting through the air is evening primrose.

The reason why the sea has this brownish hue around here is this: the Severn, Britain's longest river, drains one-twelfth of its land area –

much of the Midlands and eastern Wales, for example. So by the time it flows into the sea, it's carrying huge volumes of sediment and silt, much of which is deposited on the mudflats. These mudflats extend up to half a mile at low tide, providing habitats for shrimp, lugworm and ragworm – and important breeding grounds for wading birds such as oystercatcher, curlew and redshank.

Walking alongside the dunes, the tide way out, I soon see what look like giant teeth in a long jaw of a fossilised dinosaur jutting out of the mud. Entire tree trunks have been washed ashore among the pieces of driftwood and seaweed. An oystercatcher is feeding in the shallows. As the shoreline curves westwards, the hazy outline of Hinkley Point Power Station, shrouded by thin afternoon mist, becomes the dominant landscape feature. We have the dunes and the beach to ourselves for the last 2–3 miles into Burnham-on-Sea. The dog has a good sniff around the base of the nine-legged lighthouse, and as the dunes peter out we're soon sharing the vast shore with a few dog walkers and joggers.

Burnham-on-Sea is thronging with promenaders of all ages, and most of the seafront benches are occupied. The shingle mound of Stert Island is visible, an important sanctuary for seabirds. There is a children's paddling pool on the beach and a pavilion on stilts. Further down the promenade is a stall offering jellied eels, cockles, mussels and fresh crab sandwiches. Our immediate concern is to find a suitably robust replacement for the scruffy-looking baler twine attached to the dog's collar, so we walk into the town centre and find an RSPCA charity shop. 'Oh I'm sure we can find something better than that for her,' one of the ladies serving behind the counter assures me. She disappears to rummage through some boxes at the back, the dog meanwhile lapping up the attention of the other volunteer, and comes back with a length of canvas luggage strap, in bright peacock blue. Having deftly improvised a loop at one end for me to stick my hand through, she fixes the buckle end carefully to Fly's collar. 'There – you can join the Conservative Party now!' she tells the dog, stroking her head.

'You've done this before, I can tell,' I smile. A bargain at 50p.

At the tourist information centre, I ask where the old railway station used to be and whether there's anything left of it. The helpful woman assistant directs me over the road towards Somerfields and a play centre, but there's something else on my mind. 'By the way – what are those odd things poking up through the sands like rotten teeth?'

It turns out these blackened timbers are what remains of the Norwegian barque *Nornen*. Ferocious gales whipped up the sea in early March 1897, and the SS *Nornen* was being blown towards Berrow Flats. In the nick of time Burnham's lifeboat crew managed to come alongside the stricken vessel and rescue all ten hands on board – and the ship's dog. The shifting sands soon buried the wreck, although it was revealed again in the 1950s. Generations of children have since enjoyed clambering among the stone-hard timbers.

The chap staffing the play centre in Pier Street, which used to house the lifeboat, shows me the only visible remains of Burnham's railway station – a few feet of platform edge running alongside the lower side wall of the play centre. 'The rails used to run down Pier Street and connect with the jetty,' he says, pointing out some old photos on the walls of the play centre. 'Goods could then be transferred from train to ship. Post-Beeching they closed the whole Somerset and Dorset Railway, not just Burnham station. When you think that people once sailed to New York from here, you do wonder – what is there here now?'

Technically, Burnham still has a railway station. The trouble is that it's 2 miles away from the town centre in Highbridge, on the main line. We head south, past the Burnham-on-Sea Holiday Village, to meet the mouth of the River Brue and follow its meander on an elevated wide tarmac path. An abundance of hedgerow birds are feeding on many clumps of thistle. What looks on the map like it might just be a footbridge over the river turns out not to be, so we have to press on towards the main road. The Highbridge livestock market is in full swing, several sheep to a pen, awaiting their fate, prospective buyers huddled in conversation.

Approaching West Huntspill, a woman on the far side of the field waves at me and shouts, 'Can you help me please?' I go over to her.

She's got a sleeping baby in a buggy, and a girl of five or six in tow. 'I can't get the buggy through the kissing gate. Could you help me get the buggy over the fence?'

'Of course,' I reply, although we try and fail to get it over the barbed wire fence with the slumbering infant in situ. I suggest she takes the baby out and I'll pop the empty buggy over the fence to her. My good deed for the day done, we agree there ought to be more spring-action gates that close automatically. Kissing gates and stiles are not exactly family-friendly for walkers with babies, she says.

Along Church Road into Sloway Lane. Past an old parish water pump and pond, then left towards Stretcholt. A flotilla of canoeists is paddling up the unusually straight Huntspill river. It's an artificial watercourse, I discover, dug in 1940 primarily as a water supply reservoir for the Royal Ordnance factory at Puriton. Nowadays its main function is flood alleviation.

Having steep banks and little marginal vegetation, the river was not particularly conducive to wildlife in its early days. Since the Environment Agency took over responsibility for it in 1999, the area is now an important habitat for wintering waterfowl including shelduck, goosander and Bewick swans, and a foraging area for wading birds when marginal mud is exposed. Snipe, redshank and heron are all regular visitors. Newly planted reeds and sedges provide an important habitat for dragonflies, including the rare hairy dragonfly, and the southern hawker and emperor. Further along the lane there's some young willow in the hedgerows, the first I've seen so far in Somerset. Near Lethbridge Farm, some excellent views over towards Hinkley Point, now more clearly visible, open up.

At Pawlett, we descend on to the path adjoining the eastern banks of the River Parrett, a herd of cattle providing fleeting interest. There doesn't seem to be much evidence of planting along the banks of the river at this point. Apart from a skinny-looking fox, there's no sign of any wildlife, not even any birds. Past Bridgwater Business Park, we pick our way through Dunball Wharf where there are great piles of what look like kerbstones. Now you have to keep your wits about you here, for the M5, the railway mainline and the A38 run almost cheek by jowl, north–south, for a few miles. And nowhere do they become more intimately bunched together than at this area between the Wharf and Puriton. No wonder the wildlife has scarpered.

Having crossed the A38 near King Sedgemoor Drain, we pick up Station Road heading towards the railway line. Once safely across the railway the next task is to check out my hunch (more of a prayer really) that the footpath goes under the M5, rather than just stops by it. Sadly the footpath has been obliterated by the motorway, and I can find no evidence of it being rerouted in any way. So we have to sprint for our lives over the A39, heavy with traffic coming off the motorway, to reach the footbridge over the M5 at Down End.

As anyone who suffers from vertigo knows, attacks are unpredictable. Sometimes I can skip over a motorway footbridge without a care in the world, as in the case at Down End. Other times, I have to duck out at the last minute, like a horse refusing a jump, or manage to climb halfway up and get rooted to the spot. The dog's used to this. She just waits patiently while I sort myself out.

At the final motorway footbridge of the day, I try very hard to climb those final few steps to the top. A successful crossing would have had us on the outskirts of Bridgwater in no time at all. All I can manage is to reverse down the steps and return to the bridleway; and so we have to come into town via the Westonzoyland road.

7

The Fastest Way to Slow Down
Bridgwater to Exeter

In which we take a 'space walk' through Somerset, sprint around Watchet; and hear how a family of railway enthusiasts brought new life to a long-abandoned Devonshire station.

Bridgwater became a major port in the Middle Ages, its riverside docks filled with schooners and barges, and later the brick and tile-making trades flourished. During the Civil War, its most famous son, the Parliamentarian Admiral Robert Blake, pursued the Royalist fleet to the Mediterranean and defeated them. The Duke of Monmouth was proclaimed King here: shortly afterwards, having lost the Battle of Sedgemoor, he was executed by King James. For much of the past century, the former British Cellophane factory was a major employer.

Approaching the centre from the direction of Westonzoyland, I see estates of modern houses stretching in all directions, and in some cases, with their tiny gardens squashed up against the footway alongside the A372. The railway station looks gloomy and unloved: a lick of paint and a few hanging baskets wouldn't come amiss. Perhaps the matter's in hand.

We head towards the town centre, passing the old Bridgwater Infirmary and the River Parrett Inn, where the legend 'Karaoke with Brian – this Saturday' is chalked on a blackboard. An information board by the river bridge fills in the gaps in my knowledge about the local tides; and of the Bridgwater bore, a lesser-known cousin of the Severn bore. The tides, it says, are caused by the gravitational pull of the moon and the sun on the earth's oceans. Tides are also affected by the earth's own rotation, the shape and position of the earth's land masses. The shape and size of the Bristol Channel funnel and amplify the ebb and flow of the tides, so the distance between low and high water can be as much as 46ft; the second highest tidal range in the world. (I later discover the highest is in the Bay of Fundy in Nova Scotia.)

There are two high tides and two low tides daily. The distance between high and low tide gets progressively bigger each day, and then smaller again depending on whether the sun and the moon are pulling together or against each other. The bigger tides are called spring tides and the smaller ones are called neap tides. They occur in a rhythm that correlates with the phases of the moon. A river bore is a tidal wave caused by a strong tide pushing its way up a narrow channel with a gently rising bed, as in the Bristol Channel, and then forcing its way into the narrow river channels.

A steep-fronted wave is formed as the incoming tidal water piles up against the current of the river. Parrett Bore can reach over 2ft in height travelling upstream at about 5 miles an hour. There are only a few rivers in the world that have the right conditions to form a bore and the Parrett is one of them. For centuries, the river was the main highway in this part of Somerset; the bores and the strong incoming tides were used skilfully to carry boats miles upstream.

The tree-lined King's Square, on the site of a thirteenth-century castle, seems very pleasant, bedecked with flowers and surrounded by fine old buildings. Cornhill, the medieval market place, is most attractive, and there are plenty of people milling about in the sunshine. A campaign to restore the world's oldest concrete building, Castle House built in 1851 in Queen's Street, is being led by friends and supporters of the late Joe Strummer. The former Clash frontman, who died in 2002 aged fifty, spent the last years of his life in Bridgwater, and what's known locally as 'the concrete castle' was his favourite building.

An older gentleman I flag down for directions kindly shows me the way down to the Bridgwater and Taunton canal path. Now it may only be 15 miles long, but with a little imagination you can traverse the entire solar system in a matter of hours along this path. This is because it's a space walk too, so along this small stretch of the Somerset countryside you can become a time lord for the duration.

Beyond the industrial estates of Bridgwater, we're soon back in open country. The dog finds Frank Parsons, angling with his teenage son. 'We've seen every other type of dog this morning,' he says, 'but until now, not a collie.' By mid-morning he's already caught and netted three roach. He catches perch, tench, carp and roach hereabouts. In a few hours he has to go off and work a 2 p.m. to 10 p.m. shift at the local dairy. He's got a collie, too, at home, acquired in similar circumstances as Fly was. After a few minutes of collie chat, I tell him about where I've walked so far, my destination and the reason for it all.

Frank discloses that he used to live in the Corsham area, having worked in the military for many years. On a tip-off from a local man about twenty years ago, he went to check out a story that as dawn breaks on or about Brunel's birthday – 9 April – the rising sunbeams illuminate the entire length of the Box Tunnel. 'I couldn't say that I saw the sunlight going all the way through. It's more of a reflection because the tunnel isn't completely straight is it? It's got a slight curve to it,' says Frank. 'So it's physically impossible to see the entire length of it. But it was pretty amazing to see quite a long length of it illuminated from the Corsham entrance to the tunnel. You know there's a secret entrance there too – supposedly to the nuclear bunker. Apparently there's a massive underground site there.'

I tell him that I'd seen the portal and, from the train – once I knew what I was looking out for – what remains of the separate entrance; and read all about the military's underground operations, at the Corsham Visitor Information Centre. It was a gripping story.

'I always wondered if they would ever open it to the public,' said Frank, the dog lying flat out between us, cocking an ear to the conversation. 'I went down into the old stone quarries once and one of the underground military storage areas. Quite amazing.'

'Really – you've been down there?' I reply. 'Now, wouldn't it be great if it was opened up so people could go down there to look round. Well, I'm glad my dog introduced us!'

Having taken our leave of Frank and his son, we pass the Boat and Anchor Inn, and walk under the M5. A Virgin train flashes past us on the left, causing only a few turned heads among the slumbering cattle in the field on the right. The canal has plenty of meander, and an abundance of birds, bees and butterflies dipping and diving along the water and verdant banks. Buzzards swirl about overhead. Soon I come across a pair of swans with three young cygnets standing on the bank right next to the path. The adults hiss loudly at the sight of the dog and me. Swans can break a person's arm, someone once told me. With barbed wire fencing, nettles and brambles on the left, and the canal on the right, I am a bit vexed as to what best to do. So I look behind me and see two walkers catching up with me. 'Would you mind if I walk with you in formation past these swans?' I ask. 'Safety in numbers and all that.'

'It's the dog they're probably concerned about,' said the older man. Confusingly, they're both called Peter. We're walking roughly parallel with the railway and the River Parrett, passing red-brick Second World War pillboxes at regular intervals, and the planet Uranus waterside art work. 'Actually, I don't think the word planet means "happening" as is stated on one of the information points back there,' says Peter the Younger. 'I think planet means "wanderer". The Ancient Greeks looked at the planets and could see they were moving, while the stars remained stationary.'

'So they were wanderers,' I reply. Wandering off on another trajectory entirely, I add, 'Did you know that the planet Uranus was discovered by Sir William Herschel in Bath before he set up his observatory in Slough? I saw the observatory on my walk, or rather the site where it used to stand. It's all offices now.'

The Somerset space walk is a scaled model of our solar system, with the planets spaced along the canal. It uses the same scale for the planets and the distances between them. The sun is at the canal centre, the halfway point, and the planets are shown in their orbit in each direction, so you can see Pluto, as it were, in both Bridgwater and Taunton.

'Do you know we've got our own bore in Bridgwater?' interjects Peter the Elder, adding that he had never actually witnessed the bore personally. Without warning, the conversation lurches randomly again to alight on the nasty smell that used to hang over the town from the former British Cellophane factory, specifically from its 'stinking chimney'. From as far back as the 1930s, the town was legendary for its unpleasant odour. Of what, I ask? 'Well, farts, basically,' says PTE.

'Oh dear, that's not very nice is it, having to put up with all that!'

'Seriously,' continues PTE, 'when it was very, very strong, it did catch the back of your throat and give you a slight burning effect.'

The smell has now gone, along with 250 jobs at the factory, and it's been mooted that a wood-burning power station might be built on the site. Will this be preferable to the stinking chimney, I wonder. 'It went to a public inquiry, but I'm not sure what the outcome is,' says PTY. Paradoxically, what causes most anxiety locally about this supposedly environmentally friendly form of power generation is the fear of unacceptable pollution – from the lorries rumbling in and out of town bringing in supplies of wood.

'It was the same story at Cricklade near where I live,' I tell the Peters. 'That went to a public inquiry a few years back and the planning inspector threw it out, after local opposition to the prospect of lorries going in and out all the time with the wood. I suppose it's legitimate concern really. Life's very complex . . .'

PTE agrees. 'As we haven't got the stink of British Cellophane any more, I thought to myself – it's possible that the value of our houses might go up! Then when this plan for a wood-burning plant emerged . . .'

'. . . you became a NIMBY, did you?'

'Having always been vehemently anti-NIMBY, I suddenly realised I'm one now!' PTE admits. 'But I had thought it could be the beginning of a renaissance for Bridgwater. The factory closes down, putting something sensible in its place.'

We've reached another pillbox, and soon Fordgate Farm. The Peters are turning off the path here, going over the railway line to continue along the River Parrett trail, and then return to Bridgwater, but not before recommending to me the Brick and Tile Museum. Shame. I just don't get the same sort of high-quality conversations with the dog. 'Good luck with the rest of your walk,' says PTE. 'There's a café a bit further on.'

'Thanks – enjoy the rest of your day.'

* * *

Fly and I rest up for a while at Standards Lock, and then before you know it we've reached Saturn. It's the first thing to look for through a telescope apparently, the second largest planet with a density less than water. Magnificent rings of ice and rocks. One moon, Titan. Can it support life? The dog wolfs down a bit of cheese and pickle sandwich, having posed – loyally guarding our vital supplies – for a picture by the lock. Soon – a sign. Taunton 14.9km (9.2 miles). Bridgwater 8.6km (5.3 miles). In urban areas, the proliferation of signs has become a public menace, cluttering up otherwise pleasant views of streets and buildings to the point where they are much resented and hardly ever read. Out in the sticks, I read every one.

It's so pleasantly relaxing here, the temptation to lie down, close my eyes and just listen to the birds and the trains for a while is almost overwhelming. This canal is fantastic, alive with recent planting and greenery, abuzz with insects and waterfowl. Water lilies with yellow flowers just about to burst forth. Left and ahead there are yet more gently rolling hills. Not knowing my Quantocks from my Mendips, and finding the map unhelpful on the point, I make a mental note to find out which ones are where.

Travelling long distances with a dog at your heel can be both a blessing and a curse. Dogs will sniff out the right path, while you're still route-finding, wrestling with a damp and disintegrating map, trying to turn it the right way round. They find interesting people for you to talk to or walk with for a bit, good spots to take a break and dispose of cheese and pickle sandwich remnants, for example. Conversely, mention of a dog when you're trying to fix overnight accommodation on the phone often brings the conversation to an abrupt halt. Cosy rooms with comfortable beds and warm welcomes are promised on endless B&B websites, sometimes accompanied by long lists of gushing citations from satisfied guests. A frosty silence generally descends when you get on the phone and confess that you are travelling with a dog.

'Jupiter: the big one. One thousand three hundred and twenty-three times the mass of Earth, a dense complex atmosphere posing many questions. Weather in separate rings, enormous winds, one vast storm in space, with its giant red spots and 16 strange moons, Jupiter needs a book not a plaque.' Indeed.

I quicken my pace as we approach the café at the Maunsel Canal Centre, which has a superbly painted mural covering one side of the building. Somerset seems to excel at encouraging murals on otherwise unremarkable public buildings. There's an odd little contraption on the wall that promises an audio guide to the canal, if you turn the handle on a little wheel at exactly the right speed. How very innovative! I spend about five minutes turning the handle very slowly, quite slowly, moderately fast and finally hell for leather. At no point are any sounds being emitted from the speaker. Well, at least I'm going to get a nice cup of tea here. It being a Thursday, the café is unfortunately closed and there's no one about to ply with questions. Surely there's something one can do at the Maunsel Canal Centre on a Thursday. I read the information board and look at the last remaining cottage on the canal.

The Bridgwater and Taunton Canal is quite short at just 14½ miles. Opened in 1827, its primary purpose was to bring coal to Taunton from Bridgwater docks. Originally the intention was to link the canal into a larger network of waterways from Bristol down to Exeter. However, as the railway took over much of the potential traffic, the scheme was abandoned. Having fallen into disuse from the Second World War onwards, the canal was taken over by the British Waterways Board in 1963 and a programme of restoration was started in collaboration with the county and district councils – completed in 1992. The restoration has involved managing the clean wetland area, and restocking the water with carp, perch, pike, bream and tench. The rich food supply attracts kingfishers, herons, and in spring and summer the banks are alive with butterflies, dragonflies and crickets. The tow-path forms part of the Sustrans national cycle route 3 from Bath to Padstow.

Unusually, this canal information board actually explains how locks work. How do you move a boat uphill? On this canal, the paddles are lifted with the help of a heavy metal counterweight, hung on a chain. The boat enters the lock chamber through the open lower-level gates. The gates are closed and then water flushes into the chamber via a sluice or paddle. When the water levels have equalised, the upper-level lock gates open and, hey presto, the boat can move out of the lock and continue on its way.

Near Firepool Lock, the canal passes what was once a GWR locomotive depot. The adjacent water tower built over two lias stone limekilns provided water for the steam locomotives. In 1866, the railway company bought the canal for £64,000. The canal declined and commercial traffic ended around the turn of the century. Just past the Old Engine House (1826) and Engine House Cottage, a dredger with two crew aboard comes chuntering into view, scooping up great swathes of green gunge from the canal bed. Firepool, on the north-eastern edge of Taunton, provides a junction between the canal and the River Tone, and contains the Children's Wood – one tree planted for every child born in the borough of Taunton Deane in the year following their Environment Week in 1992. Volunteers have since planted a further 4,000 small trees and shrubs along the site.

Approaching the centre of Taunton, near the Somerset Cricket Ground, I ask a man walking towards the supermarket for directions, and wonder why he's looking so fed up. He tells me he's just watched Somerset being given a real trouncing by Durham, and Somerset are having to follow on. Does that mean the match is continued tomorrow? No, he says, it means the match carries on this evening. Cricket is such a mystery.

I find my way through town to Blorenge House in Staplegrove Road, where I've managed to book B&B for self and dog. Over the road to the King's Arms to sit in their courtyard garden. The barmaid who takes my order makes a fuss of Fly and brings out a photo of her beloved alsatian.

* * *

Another beautiful sunny morning, not too hot, cool breeze, and there may just be time for a detour along the West Somerset Railway, if I get my act together. We taxi to Bishops Lydeard, as the bus would be cutting it too fine for catching the 9.40 train to Minehead. At 20 miles long, with ten stations, the West Somerset line is England's longest preserved railway and passes through some of its most unspoilt countryside.

At Bishops Lydeard station, the platform is strewn with flower planters and baskets, edged with picket fencing and painted benches. Station staff and volunteers are dressed in dark suits and waistcoats, white shirts and dark ties. It might be too ambitious to travel all the way to Minehead and back to Bishops Lydeard, get a bus back into Taunton and then walk on from there to our next overnight stop to Wellington. So, at the ticket office I ask for a return to Watchet – a few stops short of Minehead – for the dog and myself. Although she is issued with her own ticket, at no point is she asked to show it. The journey is a delight, so relaxing that I doze off and nearly miss our stop.

The West Somerset Railway Company was formed in 1857, employing Brunel as its engineer. Construction work began at Crowcombe Heathfield in April 1859, by which time Brunel's health was failing. British Rail closed the line in 1971, but over a period of several years it was restored and reopened in stages by a group of dedicated volunteers supported by a small team of staff. The linesides are so rich in habitats for animals, birds and plants, the railway sometimes runs special wildlife trains with expert commentary on the sights, including glow-worms at dusk.

There are plenty of people on the diesel-hauled train, and Watchet is packed with ambling holidaymakers. Mindful of the time, I seek out highlights of the Watchet Heritage Trail at something approaching a sprint. It's useful having Fly at such times since she eclipses my attempted sprint with an excited powerful lope and thus hauls me along the narrow streets. After visiting Watchet in 1797, Samuel Taylor Coleridge – whose ghost seems to have been pursuing us relentlessly since passing through Box – wrote *The Rime of the Ancient Mariner,* and its harbour is thought to have inspired the port from which the mariner set forth on his fateful voyage. A bronze statue of the Ancient

Mariner stands on the Esplanade. Between the new Marina, opened by Robin Knox-Johnston in 2001, and West Street Beach are the station house and engine house of the former mineral railway. Along this line, trucks carrying iron ore from the Brendon Hill mines were shipped from Watchet harbour to the smelting works in South Wales.

Coleridge was an insatiable walker, and wrote much of his most memorable verse while pursuing a life of rustic simplicity at Nether Stowey in the Quantock Hills. Having moved there with his wife Sara and their young son in 1797, the poet once apparently walked from his cottage to Porlock in a single day. A recently devised walking trail – the Coleridge Way – enables ramblers to follow in the poet's footsteps along the 37-mile route, over the Quantocks, the Brendon Hills and Exmoor.

At the junction of Swain Street and Anchor Street is Watchet's Coronation Clock whose face bears the letters Elizabeth II instead of numbers. Back at the railway station, I pause by the Watchet Jubilee Geological Wall, an impressive patchwork of quarried local stone – iron ore, alabaster, red sandstone, lias limestone – topped with a capping of beach pebbles, iron slag and ammonite, and edged at the foot with a railway sleeper and a mosaic. It brings to mind Mr Brunel's exquisite stone collection that I saw in Swindon a few weeks ago.

The steam train back to Bishops Lydeard chugs along sedately through the seafront request stop at Doniford Halt, where you can see over to Steep Holm and Flat Holm to the north, and Penarth in the distance; then Williton and Stogumber before stopping for ten minutes or so at Crowcombe Heathfield. The guard, sporting a red carnation in his buttonhole, invites us all to step outside here and stretch our legs, and soak up the atmosphere of perhaps the prettiest station along the line. The flowerbeds are immaculately maintained. Further along the platform small children are hoisted up – like it or not – on to the locomotive and into the arms of a driver, who looks like Methuselah, in a black railwayman's cap, his grey beard extending down to his waist. Old advertisements for Bovril and Brooke Bond tea are screwed on to the station walls. Milk churns and litterbins crafted from reclaimed wood stand to attention along the platforms. It conjures up images of a bygone era of rustic simplicity and innocent pleasure. Surely any minute now apple-cheeked children wearing sailor suits will emerge from clouds of steam and smoke to wave pressed cotton hankies at the departing visitors.

All these preserved railways lay on stacks of events for children – Thomas the Tank days, Santa specials and so on. My suspicion is that these are tolerated well enough by children, subject to acceptable bribes, but often enjoyed more by accompanying parents and grandparents. In my carriage, on the way back to Bishops Lydeard, I sit near a family of four. The adults occasionally break the silence by reading out interesting snippets of information to their assembled brood from a stack of leaflets, scattered across the table. 'There's a Bakelite Museum at Williton,' says the woman to her husband. 'Really?' he sighs, staring out of the window. The elder daughter has her nose stuck into *Harry Potter and the Half-Blood Prince,* careful to affect occasional interest in what her parents have to say, in the interests of family harmony. The younger daughter squirms about in her seat, knowing that she has to try to behave herself for a bit longer, before being released back into the real world.

Back in the strangely welcome buzz of Taunton town centre, the dog and I resume our walk, heading west out of town alongside the River Tone. The warm fug of nostalgic longing for a golden age of railways that I never personally experienced takes a good hour or two to lift. A walkers' sign says we are now on the Two Counties Way, which is news to me. There was I thinking that we were on the West Deane Way – or, at the very least, on the West Country Way. There appears to be an endless proliferation of long-distance walking, cycling and riding trails, admirable in so many ways, but also slightly confusing. Somerset alone has at least seventeen, I note from a tourism leaflet, not including the Coleridge Way, nearing completion.

What sounds like a plague of locusts descending on us, a few miles out of town on the way to Wellington, is in fact a nexus of electricity pylons, the ley-lines of the modern world. The railway continues to exert a benign presence: the sounds of train hooters, whistles and sirens help me to orientate myself and confirm that we're on the right path. Just as I am feeling a little unsure of the way ahead – not lost exactly, but in danger of running out of clues as to my whereabouts – along comes a train. The sight or sound of one, whooshing out of nowhere and gone again in seconds, is sufficient to bolster confidence in my own map reading, or more importantly to undermine it. Either way these subtle signals really do help to inform the lone walker either that they are on track – or have missed a turning and are consequently roughly between the Middle of Nowhere and the Back of Beyond.

On top of the wooded hill, the Wellington Monument pierces a grey, overcast sky. It's a bit of a mystery why Arthur Wellesley took the title of Viscount Wellington of Wellington and Talavara. Better known as the Duke of Wellington, he is thought to have visited the town of Wellington, Somerset, only once, despite having an estate in the area. On the highest point of the Blackdown Hills, the first stone for the obelisk-like monument was laid in 1817 to celebrate the Duke's victory in the Battle of Waterloo. The 175ft-tall structure was completed in 1854, and apparently contains an internal staircase up to a viewing platform. The Duke famously disapproved of the new railways on the grounds that they 'encouraged the common people to move about needlessly'. When forced to travel by train he would apparently sit alone in his own personal carriage, well away from the hoi polloi.

The footpaths all the way from Taunton to Wellington are excellent, the fields devoid of cattle, and I can't believe my luck. We walk on a pleasing mixture of mown grass and straw stubble. Between Bradford-on-Tone and the railway crossing, we come across a curious sign, which contradicts all other request signs relating to livestock, their young and the presence of dogs I've ever seen. This one says: 'Beware – Cows with young calves in these fields. It is advisable NOT to have dogs on leads in a field with cattle.' Shortly I see the first sign for the West Deane Way, and others informing walkers they are on a Parish Circular Walk. Perhaps it's just best to rely on the OS map – and that's precisely what I do.

Making our way through Nynehead, we soon have some musical accompaniment. Through the open doors of Nynehead Village Hall (voted Village Hall of the year in 2003), I glimpse a pair of elderly ladies fox-trotting to the tune of 'What a Wonderful World'. They're very civilised in Nynehead.

I'd hoped that we would arrive at Nynehead Court in time to view the gardens, which are open to the public, but they're just about to close for the day. Instead I peer into the dark recess of the restored medieval Ice House, just off the right of way, and follow the path round the side and back of the property, then across a field down to the River Tone. Once over the buried remains of the Grand Western Canal and across the railway line we head into Wellington. We walk through the housing estates and past the entrance to the Relyon bed factory ('The best beds in the world') in search of the Blue Mantle Hotel.

In the town centre, a very grumpy motorist hurtles round a bend, hammers his horn, bares his teeth at me and jabs a forefinger at the traffic lights. Goodness me: someone's got out of bed the wrong side this morning. It would appear that I have stepped into a clear traffic-free road to reach the other side of Fore Street nanoseconds after the pedestrian lights had turned red, obviously a hanging offence in Wellington. We get a warmer reception from the couple who run the Blue Mantle and from their elderly, very large boxer dog Norman, stirring himself from a nice snooze on the hall carpet. Unusually for a boxer, Norman has an impressively long tail, and isn't mad.

* * *

There can't be that many municipal parks that can boast their own haha. Hahas are more frequently found on estate land, usually to keep one's livestock and deer away from one's formal gardens, while at the same time maintaining an unbroken view of the surrounding landscape from one's residence. Interestingly, Wellington Park's fine red sandstone haha provides a barrier between the formally laid gardens and a large recreation area of rough cut grass and a half-pipe of a skatepark. Donated to the town by the Fox family and recently restored with the help of a National Lottery grant, the park looks wonderful this morning, with well-tended flowerbeds and plenty of seating and shelters. Shame about the litter and bottles scattered about near the entrance off Mantle Street where I came across a group of teenagers knocking back beer and wine last night.

Underneath the railway line via minor road. Two women riders are unsure when I ask them where the Grand Western Canal path starts. 'Rockwell Green?' suggests one, clearly making a stab in the dark.

'I've passed through Rockwell Green and couldn't see it there,' I reply. However, they are able to confirm that I am going the right way for Chitterwell, broadly in the direction of the canal path. The cleaning lady at the Blue Mantle had said she always wanted to take her family to see this canal but never knew how to get there. It wasn't clear from the leaflets she'd seen mentioning it.

About half a mile south of Chitterwell, the railway line passes Whiteball. Now it was near this spot that the *City of Truro*, built at Swindon in 1903, achieved a record-breaking speed of 102mph one year later, hauling the 'Ocean Mail Express' from Plymouth to London: the very locomotive I would have seen at the Cholsey and Wallingford railway, had I called a week earlier.

At a crossroads, a little way west of the Greenham telephone exchange, we pass a 'Welcome to Devon' sign. How nice it is (truly!) to be made welcome to Devon, albeit by a road sign. Just after skirting the south edge of Beacon Hill, a cyclist shoots out from a path leading into woods next to a stretch of water, wishing me a good morning. At last we've finally found the Grand Western Canal Country Park, and path also, it appears – part of Sustrans Route 3. I wonder whether, at some point, Ordnance Survey Explorer maps will routinely indicate the major cycle routes so that people don't have to scour shops and tourist information centres – not always open when you need them – to find out where they are.

The dog runs ahead enjoying yet more new smells, excited to be off the lead, returning to weave behind me, looking for sticks to present to me. Unusually for a canal, the water looks crystal clear. All the way along to Tiverton, the path is humming with dragonflies and bees darting in and out of the wild flowers, creating a heavenly walk. There's not a soul about; like the Duke of Wellington's railway carriage, it all seems to have been created for our own personal use and pleasure. Water lilies coming into yellow flower are dotted about all along the surface of the canal. Every bridge you pass under has a nameplate – which means it's easy to keep track of where exactly we are. Near Sampford Peverall there's a turn-off for Tiverton Parkway railway station, about half a mile away. I forget to look out for the chap with the bike.

The sad thing about some of these canals, as with numerous railway branch lines, is that the achievement never quite matches the original vision of the promoters. The idea behind the Grand Western Canal was to provide a waterway link between the Bristol and English channels. Work began in 1810, but only the section between Tiverton and Taunton was completed. It was used for transporting limestone from local quarries to Tiverton where it was processed at the canal basin in the limekilns. Rapidly superceded by the railway, the section north-east of Loudwells was already closed by 1869. The remainder continued to be used for conveying stone until 1924. The present length of 11⅓ miles is now managed as a country park by Devon County Council for recreation, conservation and education, and on today's evidence they are making a good job of it.

Approaching Tiverton, there are fine views over open countryside through the tips of reeds and bulrushes. Some of the benches have chunks of red sandstone next to them, with plaques in memoriam to children. We pass a half-ploughed field, the soil a deep coppery red. A farm worker is climbing back into a tractor cab to work through the next section of stubble. Straw bales await collection in the field to the left. A pair of walkers and a family of cyclists pass me heading towards Taunton.

The surrounding land gradually falls away to the left, at least 50ft or so, to a large valley floor, the Blackdown Hills rising up on the eastern horizon. A pair of buzzards wheel and circle one another overhead. I walk past another member of the Silent Order of Anglers, whose head turns only slightly 10 to 15 degrees north-west on hearing my footfall on the gravel path behind him. It then returns to its normal position, allowing his gaze to focus back on the mid-distance for further communion with the water, and what may – or may not – lie beneath its surface, his right hand gripping the rod even tighter, as though at any moment it might vanish.

My observations since leaving Paddington station on this walk, having variously marched or dawdled alongside many rivers and canals, are that members of the SOA often share the following characteristics:

Loads of tackle – very long rods.

Long rods often extending right across the path.

Camouflage gear.

Hate to be disturbed from their chosen occupation for the day – staring unblinkingly into the mid-distance.

Their mate parked in similar fashion 20yds along the bank.

Impassive tolerance of pouring rain – often seen grimly sheltering from same underneath a green umbrella cum tent structure.

Mute or grunt-like response to a passer-by's cheery 'Good morning!'

Occasional presence of supporting female, overseeing the catering arrangements, otherwise staring unblinkingly into the mid-distance.

We pass strawberry fields on the right, the plants covered by long ridges of white fleece, with a few discernible green leaves poking up underneath, and soon we reach the eastern outskirts of Tiverton.

During the construction of the canal, the Bishop of Exeter caused considerable problems for the canal company by refusing permission for the canal to be built within 100yds of his home. This meant that the waterway had to make a large loop at this point to avoid the rectory, now known as Tidcombe Hall.

Where the canal comes to an end stands the Grand Western Canal and Country Park Centre. Here I pick up a leaflet tempting the reader to 'escape life's pressures' by taking a ride on a horse-drawn barge – the 'fastest way to slow down'. You can even hire it for your wedding reception, a disco night, or product launch. 'Public address system and 240 volts available.' That Bishop of Exeter would surely spin in his grave.

A very large dark bay horse, resplendent in his colourful barge-drawing gear, is very patiently allowing small children, hoisted in the air by their parents, to feed him tiny handfuls of freshly plucked grass. There are plenty of takers for the barge ride, sitting patiently in the vessel ready for the off.

As we need to get to Bickleigh this afternoon, there's only time for an ice cream (for me) and a drink of water (for us both) before heading off. Just as I am trying to imagine what I would actually do with 240 volts, were they to be made available to me, I'm distracted by the sound of tinkling glass. Near Tiverton Bridge, bottles and jars are being hurled into the back of a recycling lorry on its rounds of the housing estates. We pass a couple blackberrying, dropping the fruit into plastic supermarket carrier bags.

'It's going to be a good year for blackberries?' I venture.

'Yes we've had a lot of rain, and before that a lot of sun,' replies the woman, curtly.

'It's early though, isn't it – not even mid-August?' She agrees that it is early, it not yet being mid-August.

Tiverton seems a very attractive little town, with an extravagant Gothic clock tower, though no discernible signs to the Exe Valley Way. We thread our way through the Friday morning shoppers, giving way to one another politely along the narrow pavements.

'Exe Valley Way?' says a young man I ask for directions near a river bridge. 'No this isn't the Exe, it's our *other* river – the Lowman. You need to go on a bit further and turn left.'

Tiverton Tourist Information Centre is deserted when I walk in there, but within a few minutes packed out. The lady assistant is trying to hold three conversations at once, one on the phone, one with me and another with her colleague. I pay a pound for the Exe Valley Way leaflet, and she obligingly draws on a town centre map the route from here down to the start of the path. The dog suddenly retreats from a rampaging toddler and backs into a cardboard bin full of videos, thus causing them to cascade to the floor. We flee.

The section of the Exe Valley Way from Tiverton to Bickleigh looks easy enough on the map – 4 miles of gently winding river valley.

Should be a doddle. A mud-spattered walker coming off the path, heading into Tiverton, shatters my illusions. 'You'll need to be careful – it's very boggy in places,' he warns me.

'Any cattle?' I ask, warily.

'A few youngsters jumping about, but they're all right.'

The wooded sections of the path are indeed fairly treacherous and it's not long before the dog and I are caked in mud. Trying to wipe it off my boots and trousers with a handful of dock leaves just seems to make matters worse. I resign myself to the fact that for the rest of the day I will be sporting the 'Creature from the Black Lagoon' look. Similarly the dog.

* * *

When I arrive at the Devon Railway Centre at Bickleigh, Matthew Giquel is busy unloading a locomotive from a lorry ready for their Gala Weekend. He kindly invites me back early the next morning, however, for a chat and a look round. I walk along the disused railway line from Bickleigh towards Maria and Keith Gunn's bed and breakfast place at Burn, a mile or two south of Bickleigh, passing through a gate with a 'Bull in Field' sign hanging from it. It must have been a very small bull – possibly not more than 2 or 3in high, as I could see no evidence of a bigger one. The only alternative would have been to take my life in my hands (not to mention the dog's) and walk along the busy A396 Exeter road. During rush hour on a Friday evening, this struck me as inadvisable, given the absence of a pavement. The Exe Valley was once served by a little railway that branched off the main line between Barnstaple and Taunton, at Morebath, running almost due south through Tiverton to join the Bristol to Exeter railway at Stoke Canon.

Fortunately Maria and Keith are keen walkers and don't bat an eye at my dishevelled appearance, and the less than pristine state of Fly's coat and paws. We couldn't have wished for more welcoming hosts.

Keith kindly drops us back at the Devon Railway Centre the next morning for 8.30 a.m. The Cadeleigh and Bickleigh railway station opened in 1885 on the Bristol to Exeter line and closed in 1963. Station buildings were boarded up and the site was taken over by Devon County Council for use as a roads depot and tip for about thirty-four years. You can still see ingrained salt marks on the interior walls of the goods shed.

In 1997, Matthew Giquel, his family and a small group of volunteers were looking at several potential sites in and around Devon to realise their dream of creating and running their own railway centre. After leaving school, Matthew had studied transport at university and run a model railway exhibition in St Ives.

'This was by far the best site of the ones we looked at,' Matthew tells me. 'The buildings were derelict but largely intact. We laid a narrow-gauge track, a miniature railway line and opened to the public in 1998.' Having replaced the broken windows and carried out essential building repairs, Matthew moved into the stationmaster's house, where he still lives with his wife and, now, their baby. Over the years they have acquired locomotives, carriages, model railway layouts, and added on more features as income and voluntary input have allowed. The northern end of the site, formerly the rubbish tip, has been turned into a wildlife area with ponds. Since opening, Matthew reckons they have had more than 250,000 visitors, some of whom return every year.

Although he is the only full-timer, the centre has a core of fifteen to twenty loyal volunteers who help with everything from gardening and maintenance to running special events. It's run largely by a family of enthusiasts (Matthew's parents live only 5 miles away and are actively involved) for other families and enthusiasts.

The museum, housed in the goods shed, has photos of First World War soldiers loading ammunition. Thousands of petrol- and diesel-powered locomotives supplied the front lines of the conflict along narrow-gauge tracks. In civilian life, locomotives now permanently housed at the centre, or visiting on loan, also serviced quarries and sewage works, and it is their historical role as industrial workhorses that is highlighted among the centre's displays. As and when they can, the team plan to reconstruct the original station signal-box, mostly demolished under Devon County Council's auspices, and create a model village. 'We've got the skills here to do these things – it's just a matter of when.'

A dapper little steam locomotive called *Peter Pan* is one of this Gala Weekend's star attractions. Graham Morris has already spent a couple of hours getting *Peter Pan* 'in steam'. Dabbing Brasso on a cloth, he is now giving the brasswork a final polish before the gates open at 10.30 a.m. The green paintwork is already shining, and spotless.

Back in the 1950s, *Peter Pan* was one of a trio of locomotives that hauled loads of stone at an old quarry north of Tavistock. The quarry, owned by Devon County Council, was a major source of stone used for roadbuilding and employed seventy men. Most of these were labourers, swinging sledgehammers at huge boulders to break them up into pieces that were small enough to handle and load. *Pixie* and *Lorna Doone* were the names of the other locomotives working the quarry, where almost every mechanical operation was steam-powered.

When Graham and two fellow enthusiasts bought *Peter Pan* in 1975, for £2,700, it had lain in bits and pieces, in various parts of the country, for many years. Three previous owners had bought or otherwise acquired the collection of parts, with the admirable intention of restoring and reassembling the locomotive. In each case, the planned reconstruction never materialised.

'We bought it after seeing it advertised in *Exchange & Mart*, by a chap in Leighton Buzzard. He had made a start on putting the locomotive back together. But after his wife got interested in buying a boat, he didn't get any further with it,' Graham recalls.

'Did you have the know-how to put it back together then?' I ask.

'Not entirely!' he laughs. 'We went to see *Lorna Doone*, which by then was at a museum in Birmingham, to help us work out how we might do it.' Graham eventually bought out the shares of his co-owners and decided to go it alone with the restoration. The result is an immaculate, fully functioning locomotive, enjoying a new lease of life hauling carriages of visitors around places like the Devon Railway Centre. On the open market, Graham reckons *Peter Pan* might change hands for around £50,000, although he says he would never sell.

Both Graham and Matthew have very astutely managed to turn their passionate devotion to a hobby into a way of earning a living. While Matthew now runs what's become a major tourist attraction, Graham designs boilers for steam locomotives, and carries out safety inspections on them for private clients. Incidentally, *Pixie*, the third steam engine that worked the quarry near Tavistock more than half a century ago, has also been restored to its former glory – and sits poised for duty today a few yards down the track from *Peter Pan*.

Had I not needed to walk to Exeter today, I'd have liked to have stayed longer. We bid each other farewell, after Graham and Mathew indulge my request to take some photos of them with their beloved machines.

After waiting for a gap in the traffic whizzing along the A396 and then crossing the river, the dog and I pass the Fisherman's Cot pub, some thatched whitewashed cottages and the splendid Bickleigh Castle. We've rejoined the Exe Valley Way, and soon begin the long steady climb over the hill towards Thorveton. It's a narrow single-track sunken lane, flanked by high hedgerows of hawthorn, bramble, beech, foxgloves and rampant honeysuckle. I imagine these must date back a few centuries, given the variety of species and the gnarled tree stumps that peep through the greenery every so often.

From the summit of the hill, nearly 500ft above the River Exe, there are sumptuous far-reaching views of villages and hamlets towards Cullompton, and, closer to where I'm standing, I can pick out the garden decking of Maria and Keith Gunn's house at Burn. Along this narrow lane between Bickleigh and Thorveton, the dog and I need to press ourselves into the hedgerow only a couple of times to allow vehicles to pass. Essentially, the lane just serves the few farms between the villages, as the rest of the traffic seems to use the minor roads that peel off the A3072 Crediton road.

Thorveton is a pretty little place, with a neat triangle of mown grass, a flagpole and benches running down to a brook flowing underneath a bridge. The perfect lunch stop. Parked in some of the cottage porches are brown wheely bins with 'Mid Devon District Council – Where People Matter' stamped in white lettering on their sides.

Having moored the dog outside the village stores, I pop inside to assemble the ingredients of my picnic. Here, the Saturday morning talk is of maggots, specifically maggot-infested wheely bins.

'They're all over the inside of the lid – everywhere,' says a customer to the woman serving behind the counter. 'What are you meant to do with them?'

'Have you tried a high-pressure hose, or any kind of hose, to wash them away?' I interject, picking a sandwich from the chilled cabinet against the back wall.

'Away where?' she counters, crossly. 'I'll just end up with puddles full of maggots on the ground as well as in the bin. Where will those maggots go then?' I had no ready answer.

Apparently, Mid Devon District Council has recently issued local householders with an additional (brown) wheely bin and asked them to deposit in the bin all compostable material: cooked leftovers, raw vegetable peelings, garden waste, cardboard, all in together. All organic domestic waste, Mid Devon District Council wants. Fair enough. Central government has imposed new duties on local authorities to increase the amount of recycled or composted waste, so that the proportion of waste dumped in landfill sites decreases. Luckily, all cooked leftovers in my household end up inside the dog, so I don't expect maggot infestations, should a brown wheely bin ever come my way. Perhaps councils should consider issuing every household with a dog as part of their recycling strategies.

I was planning to have my picnic on one of the benches by the flagpole and the clear water of the babbling brook, but all this talk of maggots has ruined my appetite. We settle down by the brook anyway for a few minutes. Fly recycles the pork and apple sauce sandwiches for which I have just forked out £1.80, smacking her lips at this unexpected midday bonus. We get on our way again, past Thorveton's converted railway station (now a private house called Beeching's Way), back on to the Exe Valley Way, and head south towards Brampford Speke.

Every time I spy a herd of cattle on the horizon ahead, I do a quick calculation based on a close examination of the map, and the visual evidence of the real world it seeks to represent. A flow chart of questions forms in my mind's eye. Is the right of way going to take me into this square centimetre or two of OS map that appears to be a field full of cattle? If yes, what is the age/sex/mood/behaviour/spacing/position of the cattle? Does this information suggest that they will completely ignore us as we pass through, and continue to graze or chew the cud without further ado? Or is it just possible that they will decide to have sport with us, and chase or encircle us for a round of the traditional rural game of Bait the Townie Rambler? As all country-dwellers, including livestock, are aware, no one voluntarily walks anywhere in the countryside unless they are on agricultural, hunting, shooting or fishing business. By a process of elimination, everyone else defying such categories is therefore a Townie Rambler – like me. On this occasion, en route to Brampford Speke, I can soon make out a barbed wire fence. It looks as though this will serve the very useful function of keeping the cattle and ourselves apart. Despite its capacity to shred your clothes and, on occasion, your flesh, while passing through, over or under it, a barbed wire fence is sometimes a welcome sight.

Great swathes of pink-flowered Himalayan balsam smother the river bank as it snakes its way almost all the way to the northern outskirts of Exeter. Having spectacularly failed in many areas of the UK to halt the march of wretched Japanese knotweed, we've now got this other alien invader, *impatiens glandulifera,* to give it its proper name, colonising the place, killing off many of our native species.

Coming into Exeter, and feeling pretty hot and weary, we pass the leafy entrance roads to the university buildings. A sign on a side wall of the Great Western Hotel is missing an 'E', so it reads 'GREAT WEST RN HOTEL.' At Exeter St David's railway station, awaiting a train to Dawlish to my overnight stop, a teenage boy and his white-haired granny come and sit next to me on a platform bench.

'Fucking trains – why do they have to be so fucking late all the fucking time?' the boy asks his gran, slapping the bench with his hands. 'I'm gonna have a fag. Gissa match, Gran.' 'Sixteen-forty,' says his gran, scrutinising a timetable in one hand, and rummaging in her handbag for matches with the other. 'We should have got 'ere at sixteen-forty. What's sixteen-forty? Ring your dad and say we're only just in Exeter, and we're getting on the eighteen-ten.' The boy flicks open his mobile.

The minutes tick by and the three of us watch two armed policemen pacing the platform opposite, the dog having sensibly withdrawn underneath our bench and curled up behind my heels. Momentarily turning his attention to me, leaning against my backpack, the boy elbows his gran in the ribs, and nods in my direction: 'D'you think that lady's a suicide bomber, Gran?' he whispers, loudly. The pair rock back and forth on the bench, helpless with silent laughter. The policemen maintain their watchful presence, occasionally glancing in our direction and exchanging a few words with waiting travellers.

8

Atmospheric Pressure on the
English Riviera
Exeter to Buckfastleigh

In which the author and dog become marooned at Sprey Point, discover 'Little Swindon', and wade through the River Lemon; the dog, particularly, is relieved to find the Harmony Hotel.

Despite the best efforts of Isabella de Fortibus, Countess of Devon, in the thirteenth century, Exeter became a successful port – eventually. Long before the Roman invasion, the area was inhabited by Celtic tribes people known as the Dumnonii, who named the river Eisca, which means a river abundant in fish.

Having fallen out with the port authorities, for reasons that aren't altogether clear, the Countess decided to build a weir on the tidal river purely to stop boats sailing up to the centre of Exeter, as they had done for centuries. Despite a law that was meant to guarantee free passage on any English river, subsequent Earls of Devon rubbed salt deeper into the wound. They had more weirs built and, for good measure, a quay at Topsham. Boat operators then had a choice on reaching Topsham – either to unload at the quay and hand over large toll charges to the Devon family, or return their cargo to wherever they'd come from.

By 1563, the city merchants had finally got their act together and retaliated against the avaricious earls. They pooled resources to fund the creation of an artificial cut so that boats could bypass the weirs and rejoin the Exe at the centre of town. In the beginning, what became England's first ship canal was a rather lame effort, just 3ft deep and 16ft wide, and inaccessible at low tides. Although it was better than no access at all, the new waterway fell short of expectations. In 1701, the cut was enlarged to a depth of 10ft and a width of 50ft, and this had the required effect, opening up the city to free international trade for the first time in nearly 500 years. In 1844, the Bristol and Exeter Railway opened its line into Exeter, but the city corporation refused to allow a rail link with the canal basin. They maintained their line of resistance until 1867, by which time it had become irrelevant. Ocean-going steam ships had become too big to negotiate the basin, and the canal struggled to remain viable. The world had moved on, if not the aldermen of Exeter.

Exeter City Corporation had been a thorn in the side of Brunel and the GWR from the outset. After he began surveying the county in 1836, he was furious to learn of the high price the corporation was demanding for the parcel of land next to the Exe, which the railway needed for its terminus. 'Exeter is behaving in that unfair and illiberal manner which has disgusted us on the GWR,' Brunel wrote to one of his assistants. 'They forget the immense advantages they will have by our coming there and can think only to exact from us as much as they can. Treat them with a high hand. Let them know we are sure to get our Bill and that we are angry at the idea that Exeter – for whose sole benefit this line is made – should dare to offer us the slightest impediment.'

* * *

My friends Angie and Huw Weatherhead, who recently moved from Malmesbury to Dawlish, are putting us up for a couple of nights. Angie, the dog and I catch a train from Dawlish to Exeter St David's station. From there, we pick up the Exe Valley Way to walk the 15 miles or so back to Dawlish, having arranged to meet Huw and their dog Isaac for lunch at the Turf Hotel. The wide flat estuarial landscape south of the city is dazzling under the Sunday morning sunshine. Half the population of Exeter seem to be out and about making the most of the weather. The river, the Exeter Ship Canal and the railway run more or less in parallel on a south-eastern trajectory, and there's an abundance of footpaths, cycleway and dirt tracks to choose from.

The only problem along the numerous canal and river paths running south from Exeter is that too many people are trying to use the same one, at the same time. Although there is a much wider track below and alongside the elevated path on the westernmost banks that Angie and I are using, most walkers and cyclists appear to congregate on the higher, narrower one, where the best views are to be had. Few cyclists seem to ring their bell when approaching walkers from behind or otherwise alert them that they're coming. Perhaps it is regarded as cissy. It's very irritating having to jump out of the way when oncoming cyclists are bearing down on you. Worse still, not only are hundreds of people sharing the same narrow path, we all appear to be converging on the same pub.

Inevitably, there are queues for the loos, queues for both drinks bars, queues for food, even clusters of random ill-formed queues for people yet to decide which queue to join, or work up collective queuing strategies. The hotter it gets, the more evident is the British stiff upper lip. A smartly dressed middle-aged woman in front of me, inching forward to one of the drinks bars, only just manages to beat back the tears on learning from the barman that to order food, what she needed to do was join one of the two food queues – one outside for the barbecue or further along inside for other food.

The sad truth is that the Turf Hotel is just too good, and its setting sublimely beautiful. You can't reach it by road – you have to walk, cycle or, depending on the tides, sail or row to it. The hotel and its large garden occupy the southern tip of a long finger of land, bordered by mudflats and a westward curve of the Exe on one side, the footpath and the railway on the other.

There's an even larger crowd further downstream, packed into an arena and cheering loudly as we pass. Powderham Castle – the seat of the Earls of Devon – is hosting a medieval jousting tournament, and by the sound of it the protagonists' performance is going down a treat with the assembled blood-lusting spectators. Angie, Fly and I trudge on under a blistering sun. The only bonus is that the cyclists have vanished. At Starcross, where the Exe Valley Way becomes the South Devon section of the South West Coastal Path, we seek refuge at the Atmospheric Railway Inn, where photographs and memorabilia of Brunel's so-called 'Atmospheric Caper' are illuminated on the pub's otherwise gloomy walls.

Outside the pub, there's a railway station, with a footbridge over to the pier for ferries to Exmouth, and a rather odd-looking asymmetrical red-brick building ahead, with a rather squat chimney. This is the home of the Starcross Fishing and Cruising Club, formerly a museum and before that a coal depot. Originally, this building housed Brunel's pumping engine at Starcross. Against the advice of his chief locomotive superintendent Daniel Gooch, Brunel recommended to the GWR board that the South Devon Railway, a 52-mile planned extension of the line from Exeter to Plymouth, should be powered by atmospheric pressure. In the event, the much-troubled atmospheric railway never got further than Totnes. Essentially the idea was this. You lay a pipe, containing a piston, between the rails. The piston is connected to a trolley which in turn is attached to the train. Lineside pumping stations are then used to exhaust the air from the pipe, and the resulting atmospheric pressure then propels the piston, and with it the train, forward. Ten pumping houses like the one that still stands at Starcross were built at intervals of 3 miles from Exeter. The whole project was jinxed from the start. Tests showed that the original 12in pipe laid alongside the railway track had to be replaced by one of 15in. When the first passenger trains powered by this system began running in September 1847, they often broke down and passengers in third-class carriages were called upon to get out and push.

The main problem with the atmospheric system was the decomposing effect of water and iron on the leather sealing valve which ran the entire length of the pipework. What with rats feasting regularly on the leatherwork too, the entire leak-ridden valve needed to be replaced, at a cost of £25,000, within a year. Faced with the ignominious prospect of throwing good money after bad, the GWR abandoned the atmospheric system in favour of the standard steam-powered locomotives in September 1848.

A few miles south of Starcross, holidaymakers are thronging the narrow coastal roads and paths. We pass a pub called The Sunburnt Arms, and a Spanish hacienda-style whitewashed extravagance, bedecked with red flowers, housing the reception area of a caravan park. Depending on the time of year, Dawlish Warren hosts a variety of visiting populations, human and feathered. After its railway station opened in 1905, the southern end of the Warren grew quickly into the holiday resort of caravans and chalets that continues to attract thousands of visitors in the summer.

Having passed under the railway line and traced a path through the packed funfair, we pause to look at a cluster of old GWR railway carriages, in traditional chocolate and cream livery, some with lines of washing strung between them. During the mid-1930s, the Great Western Railway's production of colourful posters showing idyllic country and coastal scenes that could be reached by train and bus was reaching its peak. Responding to the increasing popularity of outdoor pursuits, the publicity department began to publish a series of annual camping and rambling holiday guides. Never slow to spot a gap in the market, the company soon targeted the type of holidaymaker whose budget could not stretch to hotels and guest houses, but was at the same time unattracted to the idea of roughing it under canvas. As a result, more than sixty older railway carriages were converted into 'camp coaches' and relocated to sidings close to beauty spots like Dawlish Warren, to provide 'novel and inexpensive' holiday accommodation. Many of these remain dotted about the country.

The northern end of the Warren, reaching out towards Exmouth, is a vast expanse of mudflats, sand dunes and beach, which together provide nationally and internationally important wildlife habitats, protected under several national and European nature conservation designations. After the summer holidaymakers have packed up and left, the whole area slowly reverts to the tranquil safe haven for migrant birds that it has been for centuries.

On Whit Bank Holiday Saturday 1846, the first passenger train steamed into Dawlish, 2 miles south-west of the Warren. Unlike its brasher junior partner down the road, Dawlish had already established itself as a fashionable seaside resort long before the GWR reached the town. Elegant Regency villas were built along the Strand as early as 1803, and the pleasure gardens and lawn were marked out and planted. Visits from writers and poets such as Jane Austen, Charles Dickens and John Keats helped to enhance Dawlish's appeal to the discerning traveller. The inaugural rail journey took forty minutes from Exeter, only a few minutes longer than it does now, and gave Dawlish the distinction of being the first holiday resort, west of Weston-super-Mare, to be served by a GWR passenger service.

We're walking into Dawlish right next to the railway line. Here the track is shoehorned between the sea wall and the ravishing red cliffs that tower above the coastline. It's getting on for 6 p.m. and as we come level with some houses on the other side of the track, we're on the lookout for the 'Torbay Express', heading back towards Bristol Temple Meads. Ahead of us, the footbridge over the railway line is lined with photographers. All eyes are on the mouth of the tunnel through the Parson and Clerk rocks from the direction of Holcombe and Teignmouth. 'Look, here she comes,' says Angie, taking the dog's lead while I scramble about for the camera. As the 1930s GWR locomotive (4–6–0 No. 6024 *King Edward I*) makes its stately progress past us, spectators' cheers are acknowledged by passengers – and the driver – with waves and smiles.

Considering its dramatic setting, Dawlish is now an attractively unassuming little place, whose public open spaces and gardens appear to enjoy lavish care and attention. A little brook runs through the centre of town, cascading gently down a series of landscaped shallow steps into the sea. In an essay for a town website, a local estate agent David Force summed up the history of Dawlish as 'notable for its lack of exciting happenings. No famous people have been born here, no notorious murders have taken place or cataclysmic events made national headlines'.

Of course, the very allure of towns like Dawlish – the proximity of the sea – also represents perhaps the biggest threat to their future viability, not just as tourism centres but also as places to live in or commute from. Rising sea levels and increasingly extreme weather patterns are causing concern. In the storms of October 2004, nearly thirty beach huts at Dawlish were torn apart, and more than 100 people had to be rescued from trains stranded between Dawlish and Teignmouth. The sea wall suffered a relentless battering from giant waves. Some forecasters predict the sea will have claimed the entire spit of Dawlish Warren by the end of this century. For how much longer the railway can be sustained on its delightful current route, snaking its way alongside the sea wall, is already a matter of some local speculation.

Having said our goodbyes to Angie, Huw and Isaac, the dog and I set off the next morning, climbing the steps near the Parson and Clerk rocks, veering inland around the western edge of Holcombe, then down the hill to rejoin the path along the sea wall. By mid-morning the baking heat is already quite oppressive, making me feel slightly dizzy. Anticipating (correctly as it turns out) another attack of the wretched vertigo, on reaching the steps up to the sea wall, I nearly turn back to find another way to Teignmouth. What the hell, I'm here now. We ought to press on, otherwise we'll be dithering about all day. The first hundred yards or so are just about tolerable.

For most people, it's an easy-peasy pleasant stroll, and soon elderly holidaymakers using walking sticks are overtaking us. Although the horizontal surface of the wall is about 6½ft wide, there is a sheer drop on your left, and trains going past just a few feet away to the right. The dog senses my unease and responds by pulling on the lead quite recklessly towards the edge of the wall. Just what we need. Sprey Point, roughly halfway along, offers temporary distraction from my gathering anxiety and sense of inevitable impending doom. It's a narrow strip of land, shaded by trees, adjoining the path, a pleasant little spot. A few other people are arriving here, some on bikes, to sunbathe or just amble about.

Observing, with a groan, that the drop down from the sea wall to the lapping waves becomes much bigger on the next stretch into Teignmouth, I seek out the shade of some dense-looking evergreens, sit down, take a swig of water, and then pour the rest into the dog's collapsible bowl. As usual, she ignores it – preferring to seek out her own water sources while we're on the move – but you have to make the offer. A man in his late fifties, possibly older, is busy cutting and clearing some of the dead wood from the trees, making neat piles underneath the branches.

'Are these pine trees?' I ask brightly, trying to suppress rising nausea.

'Tamarisk,' he replies. 'Apparently the Edwardians brought them back from their grand tours of Asia. They liked to collect things, bring them home. This is what they brought here, planted them up. Tamarisk.' He points to some miniature versions of the adult trees, sprouting right in front of them, facing the sea. 'See these lower branches, they've grown along the ground and taken root.'

'Ah. Yes. Layering,' I say, clutching my stomach. 'Is it called layering, a type of propagation?'

'That's the word. Layering,' he says. 'Yes, the new trees form by way of layering.'

Soon, I am back on the sea wall, and completely unable to move. The drop is probably no more than 20ft or so, but to someone with an irrational fear of heights, it might as well be 2,000ft. As far as I'm concerned, we're teetering on the edge of a bottomless abyss, from which there's no return, and it's about to swallow me and the dog whole. After a few minutes of standing there like lemons, Fly gives me one of her rare reproachful looks that she saves for these occasions when my behaviour is evidently beyond the pale. Roughly translated, this look means: 'Look, you're meant to be in charge. Don't expect any bright ideas from me . . . just remember, I'm a dog.'

Eventually, my guardian angel materialises. 'Are you all right?' asks a woman striding past, surveying me through dark glasses. She's wearing a baseball cap, through which her blond hair is tied back in a ponytail. I struggle for words, and she has to lip read.

'Where to go . . . ? Furr . . . tea . . . goat? Oh, I *see*. Vertigo! Right, well, do you want to take my arm and we'll go at your pace. Would you like me to take the dog too?' It seems to take hours to reach Teignmouth, probably no more than a quarter of a mile away. Deb, as she introduces herself, is kindness personified. She manages both to keep up a constant stream of conversation with me, and convince me

The Clifton Suspension Bridge, which spans the Avon Gorge between Clifton, Bristol, and Leigh Woods, was designed by Brunel in 1829 but only completed in 1864, five years after his death.

A stained-glass window in the chapel at Tyntesfield, Somerset. The house, recently acquired by the National Trust, was most famously the home of William Gibbs, whose brother, George Henry, was a founding director of the GWR and a loyal ally and friend of Brunel.

The elegant Grade I listed Clevedon Pier on the Somerset side of the Bristol Channel. The contractors found some large quantities of Barlow rail that had been lifted from Brunel's broad-gauge South Wales Railway, which were used in the construction of the pier.

Fly beside the Bridgwater and Taunton Canal, Somerset.

A barge horse being prepared for duty at Tiverton's Grand Western Canal Centre, Devon.

A signpost to Tiverton Parkway and Willand, Grand Western Canal Path, Devon.

The 'Torbay Express', hauled by No. 6024 King Edward I, *steaming through Dawlish, Devon.*

Matthew Giquel, Devon Railway Centre, Bickleigh.

Graham Morris aboard Peter Pan, *Devon Railway Centre, Bickleigh.*

Keith Wills in the GWR Room, Newton Abbot, Devon.

The author and Fly on the South-West Coastal Path, between Dawlish and Torquay.

The Cliff Railway, Babbacombe, near Torquay.

Buckfastleigh station on the South Devon Railway.

Colin Sully, South Devon Railway, Buckfastleigh station.

Brian Thomas, South Devon Railway, Buckfastleigh station.

The Royal Albert Bridge, Saltash, Cornwall, was designed by Brunel and opened by the Prince Consort in 1859.

Martin Reed and Emma, St Germans, Cornwall.

Dave and Lizzy Stroud, and their children Walter and Poppy, outside the railway station building which they converted into their home, in their spare time, at St Germans, Cornwall.

Left: Hilary Mulley (left), Dora Ellis and Fly at the Wadebridge Concern for the Aged day centre, Cornwall.

Right: A statue of Cornishman Richard Trevithick, an early nineteenth-century pioneer of the steam locomotive, outside Camborne Public Library, Cornwall.

The author and Fly on reaching their final destination, Penzance station, which was first opened in 1853.

to believe that my phobic behaviour is not at all unusual on the sea wall. Having ensured that I was OK, she vanishes into the crowds of holidaymakers drifting along the seafront in search of relaxing, leisurely activity.

* * *

We will return to Teignmouth later, but first a detour by train into Newton Abbot, following the spectacular Teign estuary before crossing the river on the north-eastern outskirts of the town near the racecourse. On the wall of platform three at Newton Abbot railway station is a framed GWR Roll of Honour, listing the staff members, by departments, who lost their lives in the First World War. Some 2,436 employees died in the conflict, about one-tenth of those GWR men who had signed up to serve in His Majesty's Forces. About one-third of the entire GWR workforce had signed up. The citation reads: 'Many of the men were called upon to participate in some of the fiercest fighting of the campaign. They upheld the best traditions of their country and their memory is revered alike by the company and their comrades.'

The large station building we see now was built in 1927 of red Somerset brick and Portland stone. The cast ironwork supporting the platform canopies is painted blue and pale grey, with hanging baskets of artificial Michaelmas daisies, gerberas, trailing ivy and fern. Across the road from the station entrance there's a pleasant little park with a bandstand, where people are lolling about on the grass staring up at a clear blue sky. Walking down Queen Street towards the Newton Abbot Town and GWR Museum, I pass some nice old buildings that have been ruined by garish shop frontages built out on to the pavement. The museum is housed in St Paul's Road, tucked away from the busy shopping area, the street where Brunel, as engineer to the South Devon Railway, worked on his plans for the line.

The original station buildings, an assortment of 'temporary' wooden structures, were erected in 1846, and on 31 December that year, the first train was hauled into the town by the steam locomotive

Antelope, on loan from the GWR, to the cheers of large crowds and musical accompaniment from the Totnes and Teignmouth town bands. A little over a year later, the short-lived 'atmospheric' system of engine propulsion arrived, with trains running each way under this form of power between Newton and Exeter. But it was the opening of the South Devon Railway Company's locomotive works in the mid-1850s and the growth of the Cornwall Railway that really transformed Newton Abbot into a boom town.

The failure of Brunel's atmospheric railway was forgiven if not forgotten by the achievement of his last great work, some say his greatest – the spanning of the Tamar between Plymouth and Saltash by the Royal Albert Bridge. Progressively, the main line could then be driven through Cornwall all the way down to Penzance. This extension of the GWR was to the great good fortune of Newton Abbot as its railway engineering works, from 1859 on, handled virtually all the locomotive maintenance west of Exeter, and itself underwent substantial redevelopment and expansion to accommodate it.

By 1866, Newton Abbot had also become an important junction of railway lines running north-east up to Exeter via the Teign Valley; another stretching north-west to Moretonhampstead; and a third going south through Torquay and Paignton to Kingswear on the eastern side of the Dart estuary. After the broad-gauge track was pulled up and replaced with standard gauge (completed over a single weekend in May 1892), the former South Devon Railway works were substantially rebuilt. By 1901, long after the GWR had taken over the South Devon Railway, Newton Abbot's population of about 4,000 in the mid-nineteenth century had almost trebled.

'The town became known as Little Swindon, with the railway works employing around 1,000 at one time,' museum volunteer Keith Wills tells me as he shows me round the GWR room. 'My father worked in the engine shed, my grandfather in the stores and my great-grandfather was a shunter.' Keith himself worked as a schoolteacher before retirement, but from his early train-spotting days onwards has maintained a lifelong and passionate interest in the history and geography of railways.

Unlike the Swindon works, Newton Abbot's railway infrastructure, locomotives and rolling stock suffered considerable bomb damage during the Second World War. In the postwar years, as branch lines were closed and much of the rail traffic – passenger and freight – switched to roads, the Newton Abbot works fell into a period of progressive terminal decline. The locomotive works closed on 15 July 1960, and the steam engine shed a couple of years later.

The centrepiece of the GWR room in the museum comprises the lower quadrant semaphore signals from Newton Abbot station, connected to an array of levers and bells along one wall. Traffic lights replaced these signals in 1987. Museum visitors are now invited to pull and push the levers and bells so that they can see for themselves how the signals worked and how the signalmen, based in their various boxes, communicated with one another. For example, the ringing of three short bells, followed by a brief pause and another bell, was known as a 3–1. This meant: 'Will you accept my train?' If the relevant section of track was clear, the question was answered with the same 3–1 bell code, and the train was allowed to move ahead. If not, the train stayed where it was.

Keith points out various relics around the room to myself and a couple of other visitors – Mr Brunel's WC, a curious piece of coiled metal tubing that looks like it must have been a locomotive part but was actually a water heater for brewing up tea, a huge poster of the Royal Albert Bridge. There's a collection of watercolour prints by a Mr William Dawson, and these are displayed in pairs. Each pair of pictures shows the landscape on either side of the Atmospheric Railway. The museum plans to display the originals, on loan from the Institution of Civil Engineers, as part of the Brunel bicentenary celebrations. One wall of the GWR room will also be given over to a model of the Atmospheric Railway, showing how it worked.

'If the right materials had been available at the time, the atmospheric system might have worked in the long run,' says Keith. The principle, he says, is quite simple. I ask him to explain it to me simply. 'Well, say you have a tumbler of fluid, with a straw in it. If you suck the air out of the straw, then the atmospheric pressure bearing on the surface of the fluid will push it up the straw. Similarly, if you have a piston fitted inside a tube or pipe, and you suck air out of the pipe ahead of the piston, this allows atmospheric air pressure to enter the pipe from behind the piston, and this propels the train along.'

* * *

Although she was invited in, I was worried that the dog might knock over some irreplaceable relic with an ill-timed wag of the tail or some other display of canine exuberance. So I'd parked her outside the entrance to the museum with a bowl of water, and under the watchful supervision of the kindly ladies staffing the reception area. They said that she had been lying down apparently content. It would have been good to stay longer and see the rest of the museum, but I was aware of time marching on and the need for Fly and me to pick our way along the southern banks of the Teign estuary to reach Shaldon and our overnight B&B stop by early evening. So we'd better get our skates on.

Somehow, I'd managed to lose my mobile phone during the previous twenty-four hours, so it was little short of a thrill to happen across an Orange shop in Queen Street, en route to pick up the Templer Way, the footpath we're going to follow all the way to Shaldon. In a matter of ten minutes, the young lad at the Orange shop had fixed me up with a new handset and neat case, got me and it linked up with the network, and relieved me of about £75.

'Ooooh – there's a dog in 'ere,' squeals a dismembered voice from the back office of the Newton Abbot Tourist Information Centre. 'Hallo dog.' Having called Fly to order, I buy a copy of the Templer Way leaflet, as it would seem sensible to know where we're going and how we're going to get there. We dodge the traffic over the large roundabout at the end of the road, and slip down to the footpath adjoining the disused Stover Canal at Wharf Road Sidings. In the nineteenth century, locally mined Bovey Basin ball clay and Haytor granite used to be carried by barge downstream from here to the port at Teignmouth, often returning with coal, farm produce and farm animals for sale at Newton market. The canal joins the Teign a few hundred yards east of the sidings and, depending on the tides, barges would sail, drift or be poled downstream to Teignmouth. The granite was transported here from quarries at Dartmoor via a tramway built by George Templer, whose father constructed the canal. Railway sidings were built after the Second World War to handle the coal needed for the electricity generating station near Tucker's Maltings.

The dog and I wade across the River Lemon near its confluence with the Teign, as I can't face the terrifying-looking footbridge. A few quizzical glances from passers-by are cast in our direction as we scramble up the riverbank to rejoin the path. After passing underneath the A380, the footpath soon gives way to the foreshore of the estuary, and the slippery mixture of mud, shingle and large stones underfoot slows us down. Having failed so far to check the tide times, or indeed read my Templer Way leaflet, it's sheer good luck that we happen to be passing through at low tide. A canopy of trees lining the estuary banks – holly, sweet chestnut, oak, beech – shades us from the afternoon sun. Above the northern shores of the Teign, a patchwork of small fields, hedgerows and copses covers the gently rising land beyond Bishopsteignton. Behind me the glistening Teign snakes its way through the reddish-brown mud, and the Dartmoor peaks stretch across the skyline.

As we stop for a drink of water, another walker speeds past shouting: 'I'm having a pint at the pub – about ten minutes ahead.' It takes us more like half an hour to reach the Coombe Cellars pub, perched above the foreshore, with a large veranda of wooden benches and tables overlooking the estuary. It would be no hardship to while away the rest of the afternoon gazing at the wading birds on the mudbanks and the trains rumbling back and forth between Newton and Teignmouth.

By Arch Brook Bridge, a couple of birdwatchers using a telescope and binoculars tell me they've seen dunlin, godwits and curlews, and that the traffic in Torquay earlier was absolutely diabolical. Fearing a twisted ankle or dislocated paw on the greasy stones, I decide to call it a day on the Templer Way, come off the foreshore here and take our chances on the lane winding its way alongside it into the village of Shaldon.

Diane and Adrian make us welcome and comfortable at our overnight B&B stop in Fore Street, despite the two of us fetching up on their doorstep in a rather dishevelled state. The dog's coat, normally black and white, flecked with grey, has an additional reddish hue from the belly downwards, and my boots are still squelching from this morning's immersion in the River Lemon. I spread pages of today's *Guardian* across the floor in our room as best I can (thinking I might read it later on my hands and knees), laying towels over the gaps, so that Fly can pad about without leaving her signature red paw prints on the carpet.

Our evening stroll takes us through attractive narrow streets of Georgian and Regency buildings, across the green, past the bowls club, along Marine Parade, then back to the green and the Londoner's Inn just in time to watch the sun set over the rooftops of Shaldon.

* * *

Before heading south towards Torquay the next morning, we hop aboard the little ferryboat across the estuary to explore the seafront at Teignmouth, a circuitous five-minute ride between the moored fishing boats and motor cruisers. The ferryman says he quite likes this time of year – early September – when the local economy gets a late summer seasonal boost from the 'grey pound', and the fares are easier to calculate. 'During the school holidays, you get – say – two adults and three children, well, the fares take a bit of working out. It's almost all just adult fares now – a lot of older people are visiting. So doing the sums is much more straightforward!'

Twice Teignmouth has been invaded and torched by the French, first in 1340 and then in 1690. They were simply following a trend set by marauding Danes in AD 800. For three centuries until the nineteenth century, many of the menfolk of Teignmouth and Shaldon earned their livelihoods in the Newfoundland cod fishing grounds, returning home only for the late autumn and winter months. In the 1820s, Dartmoor granite was shipped from here to the capital for use in building London Bridge, the Embankment and the British Museum.

For me, the finest feature of the seafront lies just across the road from the beach and the promenade, a large open space of mown grass and neat paths known as the Den. Once a sandy waste ground used for drying fishing nets and racing horses, the Den was donated by the then Earl of Devon to the town in 1869, and then laid to lawns and gardens. This morning, there are plenty of dog walkers about throwing sticks and balls for their excited animals. At the Shaldon end of the promenade is an open-air swimming pool, devoid of water, a small hut that appears to be constructed entirely from fruit juice and milk cartons, and further along a four-sided tower on stilts covered in shells, lengths of driftwood, rope and children's plastic spades. Through the window of the hut, I can make out a notice advertising an exhibition of art work using recycled materials.

Back on the beach at Shaldon, we head up the lane towards Ness Point to join the coastal path. Climbing the first of several steepish inclines, I meet another walker. 'Do you mind if I walk with you past those cattle up ahead?' I ask, pointing at the herd of shaggy-haired beasts slumbering in the top corner of the field.

'Not at all,' he replies.

We pause briefly halfway up the hill to look back at the stunning view. In the foreground, the red cliffs rise almost vertically above Ness Cove, which can only be reached on foot via a 'smugglers' tunnel from Shaldon. In the distance we can identify Exmouth about 10 miles away on the north-eastern horizon. Geoff is staying at a hotel on the Dawlish side of Teignmouth that caters for the needs of blind and partially sighted people and their guide dogs. 'It's really very good, they've got speaking clocks, if you need them, and the menus are all in large print or Braille. They come and fetch you from the railway station and take you back there at the end of your stay.'

At Maidencombe, the dog and I make a short detour uphill, and cross a busy road to call in at Brunel Manor. In 1847, Brunel began to buy up large areas of estate land between Watcombe and Maidencombe, on which he planned to build a great mansion with landscaped gardens and staff cottages overlooking Babbacombe Bay for his retirement. He had become well acquainted with the area while working on various engineering projects in and around Newton Abbot. Over the years, he spent every spare moment planning the big picture and fine detail of his dream home. A prominent garden designer William Nesfield helped him to landscape the grounds and plant rare specimen trees, and the architect William Burns assisted with the design of the house. Brunel envisaged spending the last years of his life as a country gentleman in a Loire chateau-style residence that offered every comfort and a beautiful view from every aspect. Only the foundations were ever built. As Brunel's health deteriorated progressively from late 1858 onwards, work on the dream home slowed to a halt, the castle of his mind's eye left hovering in the air.

A paper manufacturer James Roper Crompton built the house that now stands, naming it and the estate land around it Watcombe Park. The property changed hands several times between the late nineteenth century and the Second World War. During the early 1930s, it was bought for use as a Christian Holiday Home by the Holiday Fellowship, which renamed the mansion Brunel Manor, and during the war it was occupied by the Stockwell Teacher Training College, evacuated from Bromley. The Woodland House of Prayer Trust bought it from the Holiday Fellowship in 1963, and it is now used as a hotel and conference centre for Christians and non-Christian visitors and groups.

Having asked me to sign in and wear a visitor's badge, the dark-suited receptionist at Brunel Manor kindly allows me to flick through the folder entitled 'Isambard Kingdom Brunel', a collation of cuttings, texts and photographs about his life, career and personal connection with the area. She permits me to have a little look around the public rooms on the ground floor, off the fabulous staircase. There are many photographs and items relating to Brunel's life displayed on the walls. The view from the large bay windows towards Torbay and the sea is indeed fine, although partially obscured by trees.

Mindful of the dog outside, secured to the Victorian boot-scraper beside the entrance steps, I feel obliged to return to her and get on our way. As we pass signs along the drive to Brunel Lodge and Brunel Court, all part of the centre, I feel a slight chill and quicken my pace. We pass

Rock House on our descent back to the coastal path, soon to plunge into the sunlit glades and glorious red-soil paths of the Valley of Rocks.

* * *

'Well, you can go under the cliff railway just here,' a sprightly elderly gentleman tells me, pointing at a narrow tunnel entrance halfway up the hill. 'Or you can carry on up the path alongside it. It depends what you want to do.'

'What are *you* going to do?'

'Well, I'm staying at Paignton, so I'll probably drop down to Oddicombe Beach and follow the coast path back. Such a lovely afternoon. Where are you staying?'

'Well, I've no idea. Haven't booked anything. Teignmouth Tourist Information Centre told me that there was not a bed to be had anywhere in Kingswear or Dartmouth, so I guess I'll have to try and find something in Torquay or Babbacombe.'

'You'll find somewhere – there's always somewhere to stay, if you look about.'

The two lads running the upper end of the cliff railway replenish their dogs' empty water bowl for Fly's benefit, and message down to their opposite numbers at the foot of the cliff to send up more leaflets on the next car. The cars, painted in a rather fetching yellow and turquoise, carry 250,000 passengers each season from the top of Babbacombe Cliff down to Oddicombe Beach, some 240ft below us. Sir George Newnes MP was rebuffed by the worthies of St Marychurch when he offered to build a lift up the slopes of the cliff back in 1890. Subsequent plans came to nothing until the Waygood Otis Company completed their scheme for opening in 1926, and following refurbishment after the Second World War, it was reopened in the early 1950s. Torbay Council currently runs it, and one of the children spilling out of the car that's just reached the top 'station' is already asking her parents when she can have another go.

The bus driver taking us into the centre of Torquay says he will give me a shout when we reach the best B&B quarter, well away from

the seafront and my likeliest bet for finding a place for us to spend
the night.

The dog and I find ourselves standing near Avenue Road at about
5 p.m., me gazing up and down the street, wondering which way to
go. While passing through, Charles George Harper observed in 1893
that Torquay 'pleases every variety of the querulous invalid: these
feeble folk lie here in strata, elevated or depressed, as best befits their
individual complaints.' I am beginning to suspect Charles George
Harper of being a bit of a misanthropic misery-guts.

'Are you lost?' a woman asks me.

'Just looking for somewhere for me and the dog to stay the night,' I
say brightly, trying to suppress rising gloom.

She whips out her mobile phone and speaks to a couple of her
contacts in the area. They're both full up. She points us in the right
direction, apologising for not being able to help, I thank her for
trying anyway, and then spend the next hour ringing doorbells.
Hopeless. It's the same story – no rooms or no dogs in rooms. Dogs
stay in guest cars. I don't have a car to hand – we're walking. One
hotelier, clearly anxious to get rid of us, while processing an outgoing
guest, gives me a number for the Torquay Tourist Information Centre,
but it turns out to be the number of a taxi company.

Towards the top of Avenue Road, there appears to be just one more
possibility before I cut my losses and head for the railway station. If we
have to doss down in the railway station for the night, so be it. I ring the
doorbell of the Harmony Hotel, fingers firmly crossed. We're in luck.

* * *

Before breakfast I take Fly out and around the green just beyond the
railway line and the main road. We walk with another guest who's
staying at the hotel so that she can visit her dying mother at Torbay
Hospital. The mother keeps rallying and then falling back into
unconsciousness. There's talk one day of moving her to a nursing
home, the next of keeping her where she is, very distressing for all
concerned.

I was hoping to call in at Torre Abbey to view the Agatha Christie room. She was born in Torquay and had a holiday home nearby at Greenway on the River Dart. However, the twelfth-century abbey is closed for refurbishment until 2008, so bang goes that idea.

Much of my early teenage life, at weekends, was spent curled up in my parents' caravan at Waldringfield in Suffolk, absorbing whatever Christie murder story I could lay hands on, while they went out sightseeing: *The Mysterious Affair at Styles, Murder on the Orient Express*. The more bodies the better.

We cross the main road and walk along a good wide pavement down towards Paignton, past the beach huts. The tide's in and water is lapping the edge of the sea wall below on an overcast, pleasantly warm morning. Near Oil Cove, I'm suddenly aware of the railway cutting through the cliff to my left, as the view of Paignton and its pier, and Brixham beyond, appear over the horizon.

At the railway station a decent crowd of us are waiting to experience the Paignton and Dartmouth Steam Railway. The station and platforms are immaculate, not a cigarette end in sight, and a kindly gent at the ticket office watches the dog for me while I go in the shop-cum-café. Our train is pulled into the station from Kingswear by the steam locomotive *Lydham Manor* (4–6–0 No. 7827) built in Swindon in 1950 by British Rail to a GWR design. There's a Pullman first-class observation car, the Devon Belle, immediately behind the engine, but as neither dogs (nor pushchairs nor food) are allowed in it, Fly and I rough it in one of the standard Great Western carriages – luxury of course by today's standards of cramped accommodation on modern trains.

The ride is a joy, along steep gradients, running in parallel initially with Network Rail's track until Goodrington, the giant flumes of the Quaywest Water Park and our first sight of the sea. Then we climb up the hill to Churston, and then descend into the Dart Valley. Crossing three viaducts, with steep drops of up to 95ft, and then pulling into the dramatic setting of Kingswear north of Dartmouth, is hugely enjoyable and all over much too soon. We then dawdle down to the ferry, the *Dartmouth Princess*, which takes us over the water to disembark near what used to be Dartmouth's railway station, now a restaurant. It must be one of the few stations ever built to lack the usual paraphernalia of track and trains next to it; instead, an estuary and large numbers of small boats. Painted in the familiar chocolate and cream of the GWR coaches, the Station Restaurant and Carvery seems to be doing good business. The dog finds it too crowded for comfort here, so we share a shop-bought meat pie further along the water's edge, where it's a bit quieter.

The Dart Valley Railway PLC operates an integrated public transport system, with its own buses and boats running up the Dart to Totnes, and has done so since 1973. This means that on a single ticket you can explore a large part of Torbay, the coast and inland villages and towns, without needing a car. I can't help noticing a headline in the local *Herald Express*, 'Brown bin maggot alert', about a Kingswear couple who are championing the use of biodegradable bags for the depositing of organic waste before they are dropped into the wheely bins. This apparently helps to deal with the problem of maggot infestations and smells, so hopefully news of this innovation will reach Thorverton at least.

The trouble with travelling on a preserved railway through fantastic landscape, surrounded by people like you who are doing so for the pure pleasure of it – no ringing mobile phones, or clattering of laptops – is that it is becoming a little addictive. On reaching Totnes, the yearning to have another go is overwhelming, so it's just as well the South Devon Railway is on hand to take us further up the Dart Valley to Buckfastleigh.

* * *

To reach the southern terminus of the railway, in the parish of Littlehempston, we follow the path heading south-east from the Totnes mainline station's car park and cross the footbridge over the river.

The 7 miles of line now known as the South Devon Railway was first opened in 1872 and soon extended 2½ miles to Ashburton, beneath the southern foothills of Dartmoor. The line never made a profit and was closed to passengers in 1958, and to freight four years later. A group of business and professional men, however, were convinced that a reopened steam line could be commercially viable so they formed the Dart Valley Light Railway Ltd. Backed by an association of supporters and volunteers, a collection of rolling stock was put together and the first trains began running in 1969. Lord Richard Beeching, better known for closing railways rather than opening them, performed the honours at the launch of the rejuvenated line in May of that year. It only went as far as Buckfastleigh as the section on to Ashburton was lost when the new A38 dual carriageway was built.

Three years later, the DVLR took over the Paignton to Kingswear branch from British Rail and by the early 1990s a charitable trust was formed to take over the operation of the Totnes to Buckfastleigh line, with its own separate support group of volunteers and enthusiasts.

Having arrived in time to catch the 11.30 from the Totnes terminus (now called Littlehempston Riverside), I find a seat in a 1937 'Collett' class open carriage fitted and furnished in the art deco style of the period – cube-shaped lamps, wood veneer panelling and glass in the partitions engraved with the GWR roundel. The distinctive company motif was designed by Arthur Sawyer in the GWR publicity department in 1931 and introduced throughout the network in the summer of 1934. While lacking the spectacular views of the Paignton to Kingswear line, this ride takes you through attractive scenery of woods, pasture, wild flowers and riverbanks nonetheless. In the breeding season you can sometimes see the salmon leaping upstream, so I hear.

Buckfastleigh station has a little railway museum in part of a goods shed, where pride of place is given to a steam engine called *Tiny*, the only surviving locomotive from the era of Brunel's broad gauge. Built by Messrs Sara and Co. of Plymouth in 1868 for the South Devon Railway, and featuring a tubeless vertical boiler known as a 'coffee pot', *Tiny* was employed on shunting duties for many years at Newton Abbot railway works, then as a stationary engine supplying steam to power its machinery.

Time to chat to some of the workers. Poking my head around the door of the stationmaster's office, I find Brian Thomas glued to the TV, attending closely to his secondary duties as official Test Match observer. As a train from Totnes pulls in, he announces the latest score over the public address system in his most mellifluous, gravelly John Arlott tones. The tension is mounting as the next few hours' play will determine whether England recover the Ashes. Having got off to a shaky start this morning, England are suddenly hitting sixes all over the ground.

Brian's love of railways developed while he was growing up in Pangbourne in the late 1930s and during the war years, when there were many Canadian, Polish and US soldiers based in the area and a prisoner-of-war camp on the outskirts of town. 'Once I'd travelled by train into Reading the first few times, I didn't want to go any other way, so I used to tell my mother that the buses made me feel sick,' he tells me. 'I remember the return fare for an adult from Pangbourne to Reading was 11 old pence.'

Brian has worked here as a volunteer here for thirty-nine years, after reading about plans to reopen the line in the railway press. From cleaning the loos onwards, he's done every job here apart from guard, booking office clerk and signalman. The work is shared out between about 10 full-time staff and 400 volunteers. 'We have a German signalman who spends six weeks here every year – that's his holiday – and then goes back home to Germany. Some of our maintenance staff are a little elderly – they call themselves the coffin-dodgers. They do a great job, but we're short of younger volunteers.'

Unlike Brian, who worked in health and social services before his retirement, Colin Sully went to work for British Rail Western Region at Newton Abbot straight after finishing school in 1956, cycling the 3 miles from his home in Ashburton. Having trained as a fireman he worked on his home branch line until just before it closed in 1962, and then went to work on the railways in Bristol for another sixteen years. 'The diesel trains come along and that didn't do nothing for me,' Colin tells me. 'No comparison with steam – nothing to keep your interest going. After doing shunting work for a while, I left and drove motor coaches until I retired.'

After his wife died a couple of years ago, Colin was travelling along his old branch line one day when the penny dropped, as he puts it, and he decided to return as a volunteer. 'It was like coming home, fantastic to work here again. Every day, six days a week, I'm here. I clean the toilets first thing, and then I bring out all the milk churns, wooden handcarts, luggage trolleys and cases on to the platform, help to sweep up, take tickets – whatever's necessary. Well, it's ideal for me, 'cos I'm on my own now apart from the dog, and it gives me a lot of pleasure. The price of petrol being so high, I don't go many other places. I'll carry on here as long as I can.'

The South Devon Railway markets itself as providing a great day out, not just a steam train ride, and I can see why. At Totnes there's a rare breeds centre to visit, as well as the town itself, and at the other end of the line, the chance of a ride in an open-air bus to Buckfast Abbey, or a stroll to the Butterfly Farm and Dartmoor Otter Sanctuary. They go to great lengths to recreate the charm and atmosphere of a world long gone. Outside Buckfastleigh station are parked an old Morris Traveller and an Austin Seven, belonging to members, beautifully restored.

'We like to think there's something here for the whole family,' says commercial director Neil Smith. 'It would be perhaps wrong to say that Mum and daughter like to go and see the butterflies and otters and Dad and son like to go on the railway, but people have different tastes and we try to cater for them all. Obviously sunshine helps to bring out the visitors but if it's dull or it's raining, you're under cover most of the time, and we swap the open-air bus for a covered one.'

On the train back to Totnes, Roy Skates is checking the tickets. He keeps a plastic whistle in his waistcoat pocket which turns into a snake to distract fractious children. When we stop at Staverton station, he points out a river walk or bike ride you can do from the station up to the weir, and how you can get to the village pub. Some of the railway's supporters, now well into their eighties, used the station to get themselves to school. The main structure of the station signal-box had to be rescued from a local vicar, for use on the reopened line, after he'd bought it from BR to use as a garden shed. Carpentry is Roy's specialism. He shows me some of the oak panelling he's refurbished on the carriage. 'Last winter I started on the inside of a dining car and got about half of it done, so I hope to finish it off this winter. I take the panelling out and work on it at home – outside to do the sanding, and then finish it off in one of the bedrooms. The railway supplies the varnish, but I get all the other materials myself. Takes about three weeks to do about three metres [10ft] of panelling. It's hard work, but very satisfying.'

9

Into a Cornish Autumn

Buckfastleigh to Lostwithiel

In which we walk underneath the Royal Albert Bridge; the dog is hauled along the Camel Trail in a chauffeur-pedalled chariot; we find the 'fairest of small cities' and become a little exasperated with Bodmin.

At the top of Armada Way, just below North Cross, a small huddle of us peer at a street plan of Plymouth City Centre trying to get our bearings. Some wag has turned all the fingerposts in the middle of North Cross round the wrong way – what an amusing jape that must have been.

'We're trying to find the Hoe,' a glum-looking woman tells me, turning to her utterly hacked-off-looking husband. 'You thought it might be up this way, didn't you?' They had already walked up through the shopping centre, she added – pointing towards Plymouth Hoe – but couldn't find it.

'I think you'll find it's the other side of that great big memorial – next to the Sound,' I venture. The husband doesn't look at all convinced. I leave them to it and stride down the hill, past the Copthorne Hotel, crossing Mayflower Street and Cornwall Street, around the large sundial water feature down to Dingle's department store on the corner of Royal Parade. The hard landscaping and the imaginative planting of the central reservation that parts the pedestrian traffic down to Hoe Park helps to distract attention from the austere, almost Soviet-style modern architecture.

The Hoe was used for civic memorials long before the days of Sir Francis Drake, whose statue stares out towards the Sound, left hand on his hip, a pair of compasses in his right hand, a globe at his feet. Until about 1815, it was also used for bull-baiting.

From the *Herald* newspaper office, I grab a handful of tourism leaflets and search in vain for any mention of the Royal Albert Bridge. 'If you're driving, there's a big car park and viewing area on this side of the Tamar,' a woman assistant tells me. 'Or there's a No. 1 bus from Royal Parade that goes to Saltash.'

Behind me in the bus queue is a chap in his sixties, who becomes my tour guide for the next half-hour or so. He farms just outside Saltash, and he's anxious for me to have a good look around the town and along the quay, listing several buildings of particular interest. On alighting from the bus in Lower Fore Street, Saltash Man directs me to the quay, the museum, the Guildhall and tells me not to forget Mary Newman's Cottage: 'From the water's edge you can see both bridges well – and then walk underneath them. You get another good view from Saltash railway station. There's talk of turning the old station building into a Brunel museum; and of course you can walk across the road bridge. There's a bust of the great man somewhere, you'll probably pass it. They ought to make more of the bridge and the Brunel connection, the councils round here – Bristol is so on the ball in that respect.' His wife appears, she's come to pick him up from the bus stop in their car. 'Have you got all that? That'll be ten quid please.' Saltash Man holds an upturned palm towards me, cackles with laughter and the pair of them disappear up the street.

Across the road from the Brunel Independent Mortgages centre, the GWR Railway Hotel and the Saltash Tanning Studios, the Saltash Heritage Trail is displayed on a noticeboard. It being a Tuesday and a few minutes past 12.30 p.m., both the heritage centre and the Guildhall are closed so I have to memorise the route of the trail. The bust of Brunel, meant to be almost opposite the Cornwall College learning centre, is nowhere to be seen. It's been concealed by foliage in recent times, and the plan is to cut back the bushes and re-landscape the area, to reveal and illuminate the sculpture in time for the bicentenary. Some people have argued that it should be relocated to the Waterside area beneath the railway bridge.

Saltash station is in a sorry state, long derelict and boarded up, and with trees growing out of the chimney. But if you turn left from the entrance and walk along the platform, the view opening up of the Royal Albert Bridge and the road bridge stretching away into Plymouth is astonishing. A Wessex train passes and bends round between the huge curved grey trusses of the rail bridge, Brunel's last great work, completed four months before his death in 1859. As was often the case with Brunel's projects the final specification presented an exceptionally tall order. The Admiralty insisted on a clearance height of at least 100ft between the bottom of the span and the water level at high tide to accommodate the tall-masted ships that would sail underneath it. The width of the river to be spanned was 1,100ft. The massive central pier, made of granite, had to be driven down 80ft below high tide level through mud to reach rock. The cast-iron curved trusses were floated into position and raised by hydraulic jacks 6ft per week one end at a time, to hang in the air while the piers were built underneath them.

Brunel directed the raising of the first truss from a platform built on it using flags to signal instructions to the machine operators. Although failing health prevented him from orchestrating the raising of the second one, he was able to see the finished work, carried aloft across the bridge shortly before his death. Built at a cost of £225,000, the bridge was opened by Prince Albert on 2 May 1859. Train services between Plymouth and Truro began two days later – some thirteen years after the parliamentary bill authorising the construction of a Cornish mainline railway was passed. The long-awaited new gateway to Cornwall was finally in business, later to become widely known – not always in jest – as the bridge that connects Cornwall with England.

A Brunel heritage centre, within sight of the bridge, would appear to be an ideal use for the old Saltash station building, particularly given the pivotal role he played in the development of the Cornwall Railway. A gallery, exhibition space and café are planned, so that a rolling programme of educational and cultural events can be staged here.

Before the opening of the road bridge in 1961, carrying the A38 over the Tamar estuary, Saltash was one of the busiest stations in Cornwall. Opposite the station road stands the Two Bridges pub, flying the Union Jack and the black and white Cornish flag. On the western side of old Saltash, at 9 Albert Street, stands Mary Newman's Cottage. She married Francis Drake on 4 July 1569, when she was seventeen and he twenty-four. In 1577, Mary was told that her husband had died during his voyage round the world on the *Golden Hinde*, only to be reunited with him three years later, his ship filled with treasure. Sir Francis and Lady Mary Drake became Mayor and Mayoress of Plymouth in 1580, making a new home in Buckland Abbey. She died of smallpox three years later.

I head down to the Waterside, where large numbers of swans patrol the quayside. A painting of the Union flag covers the entire front of the Union Inn, and a side elevation features a mural of a community scene. A Saltash to Plymouth ferry operates from here on certain days, and you can take boat trips around to the Hoe and the Barbican via the naval dockyard at Devonport. From this vantage point the entire construction and appearance of the rail bridge seem audacious, the grey trusses so modern-looking. It feels eerie to recall, walking underneath it, that the structure was not only erected almost 150 years ago, but remains in service.

In the car park on the Plymouth side of the Tamar, near the toll booths, is a pay-as-you-go telescope and an information board telling visitors that before the bridges were built, people either had to cross the river by ferry or take a long detour above Calstock to cross overland between Devon and Cornwall.

It's only when you're standing level with the two bridges that you appreciate how unequivocally the structure of the road crossing eclipses that of the rail bridge in terms of width – it has separate lanes for buses, pedestrians and cyclists, cars and lorries – and, more obviously, height. You can also see from here the railway viaduct a little to the west of Saltash station.

* * *

The Cornish landscape, riven with valleys, needed an awful lot of viaducts – thirty-four of them on the Plymouth to Truro section alone

– to carry the railway down to Penzance. Brunel's original designs featured fan-shaped timber supports between stone piers. As rails were laid to accommodate simultaneous Up and Down traffic, stone and brick structures progressively replaced the wooden viaducts. These were built right next to the originals to minimise disruption to services. The ivy-covered remains of original structures are still visible in places.

Walking on a footpath near St Germans station, the next one west of Saltash, the dog meets a border collie/springer spaniel cross called Emma. I fall into conversation with her owner Martin Reed, a civil servant turned carpenter, who has lived with his wife in St Germans for thirty years. Within a few minutes of helping me to get my bearings, Martin kindly offers to give me a historical guided tour of this pretty village, with a coffee stop at his house. St Germans sits on the north-western flanks of the Rame peninsula, perhaps one of the least visited but most beautiful parts of Cornwall.

'You know about the Palmerston Follies?' he asks as we head towards the station. 'These were the forts built all over the Rame peninsula and the south coast, particularly around Plymouth and the dockyards, in anticipation of a French invasion. And all this farmland round here used to be fertilised by dung from Plymouth – boats would bring it over, and limestone for burning as quicklime, for spreading on the fields and return with farm produce.'

Martin points out the main station building, now a private residence, also the nearby goods van converted for holiday accommodation, and introduces me to their owners.

St Germans railway station had stood derelict for more than a decade when Dave and Lizzy Stroud moved in with a couple of chairs and a Workmate to start converting the abandoned wreck into their home. The floors of the station, designed by Brunel in 1858, had almost rotted away; and the internal doorways were so well blocked up they had to use a pneumatic drill to make holes big enough to get from one room to the next. The couple took possession of the building in 1992 and then worked in their spare time over the next seven years to get the place as they wanted it.

It wouldn't be everyone's cup of tea, living right next to the GWR mainline with trains trundling past at regular intervals and stopping here too several times a day – especially with two young children: Walter is six, and Poppy three. Isn't it very noisy, I wonder, and lacking in privacy? Apparently not. 'We put in thick double glazing and reflective glass,' says Dave, who runs his own steel fabrication business. 'Initially, we found the freight trains were the noisiest, but the newer engines are much quieter, so we are not really aware of the noise now. The point about the reflective glass is that you can see out of the windows from inside, but people outside can't see in.'

Lizzy, a music teacher, grew up in the village from the age of four – her father was the farming manager for the Port Eliot estate. 'We love living here,' she says. ' We're near the centre of the village, have lots of space, and no close neighbours. Our parents have been very supportive, and Dave's granddad – a chippie by trade – helped us a lot with the conversion work.'

The family's hens now roam the station car park, which came with the building, and the garden is divided into two plots: one for themselves and the other – bordering the platform – for their holiday guests.

Sitting on a siding once used for loading farm produce is a converted London and South Western Railway luggage van which first entered service in 1896. 'It dawned on us that having a carriage on the siding would be rather nice and a good use of the space,' explains Lizzy. 'We bought it for £140 in 1997. A friend of ours did most of the conversion, and we were able to let it from the spring of the following year.' Stepping into the cosy van with its fine views of the mainline I can see why it's so popular with rail enthusiasts who take most of the bookings, and often become regular visitors.

The couple have since acquired two more rail vehicles for letting – a 1957 British Rail corridor carriage, which sits next to the platform at Hayle railway station in West Cornwall; and an 1889 GWR travelling post office with a clerestory roof. The latter was among the carriages being hauled by the *City of Truro* when the locomotive broke the speed record in 1904, Lizzy tells me. They plan to site it on a piece of land near their home as soon as the necessary practical arrangements can be made.

Before taking our leave, Lizzy shows me the mixed woodland she and Dave own bordering the railway line at St Germans. This supplies them with logs for their wood-burning stove, and materials for carpentry projects at home and for carriage conversion work. What an enterprising couple!

* * *

Martin Reed guides the dog and me down Gays Lane, alongside the railway line.

'See this high bank here – this is where people used to dump all their rubbish, on top of the spoil from the railway cutting. Look, here's a bit of an old china marmalade jar.' He picks up another relic sticking out of the bank, a piece of thick brown glass, and holds it up to the light to find some writing on the bottom. 'Part of an old champagne bottle I think. There, you can have that as a souvenir from St Germans!'

On we go to his house in the centre of the village, past the post office and stores, the rather austere-looking Masonic Hall, the old slaughterhouse and smithy, and the stone milk churn stand at the foot of his front garden, smooth and flat from centuries of use. On cattle market days, the auctioneer used to step up on to the milk churn stand and conduct the proceedings from there. While the coffee is brewing, my tour guide entertains me with magic tricks involving, firstly, two pieces of wood and a propeller, and secondly, two horseshoes whose ends are linked by chains. The chains, in turn, are enclosed by a metal ring. Then out come the local history books, notably *A Pictorial View of the East Cornwall Villages of St Germans, including Port Eliot, Tideford and Bolpathic 1850–1950*. This contains a photograph of Aircraftman Shaw (better known as Lawrence of Arabia), who visited in 1926 to deliver a lecture on the Priory of St Germans, an event attended by George Bernard Shaw, among others.

Correctly anticipating a conversation of some length, the dog has by now settled down in a corner of the dining room and fallen into a deep sleep.

'What do you reckon these are?' my host asks, placing in front of me on the table various bits of ancient ephemera dug up from the garden over the years – buttons, coins and two marble-sized metal balls. 'These are musket balls,' he says. 'Those buttons have come off the butlers' livery from Port Eliot, because they've got the Eliot family coat of arms embossed on them – the elephant and the coronet.'

It's all very interesting, but given the time, it would seem prudent to restore a railway theme to the conversation. 'Is there anything else you can show me or tell me about the railway?'

'Well, you ought to see Battery Cottage – that's right next to the viaduct.'

* * *

Back on the road again, and Martin is pointing out the fine parish church of St Germans, and the shafts down to the underground passageway leading off it. It was founded in about AD 430 during the visit to Cornwall of St Germanus, Bishop of Auxerre in central France. The purpose of his visit was to fight the heresy of the Celtic theologian Pelagius, who denied the existence of Original Sin and recommended free will as a philosophy for life. The present church stands in a natural amphitheatre on the site of a Saxon cathedral. Created during the reign of Edward the Elder (AD 900–25), the diocese of Cornwall was merged with that of Devon in 1040. A little over a century later, the church was acquired by a foundation of Augustinian monks who built the priory that later became Port Eliot, the home of the Eliot family and the Earls of St Germans as it remains today.

Battery Cottage stands in the most exquisite location on the banks of the River Tiddy, a short distance upstream from the massive St Germans viaduct, carrying the railway over the valley.

'See the cannons at the end of the garden? They came over from the dockyards and probably were fired to greet visitors to Port Eliot –

the seat of the Earls of St Germans – and to alert the servants at the big house to their imminent arrival. It's a picturesque spot, don't you think? I'll show you the quay – just the other side of the viaduct.' As we bid each other farewell, my tour guide recommends a lunch stop at Crafthole.

In the event, I feast on the freshest-tasting, most delicious cod, chips and mushy peas at the beach café at Seaton. In the sea here, the dog breaks the habit of a lifetime by swimming a few yards to retrieve a stick, and very pleased with herself she is too. There are plenty of people ambling about the Seaton Valley Nature Reserve on the other side of the road from the beach, and a few sitting grimly in deckchairs wrapped in towels against the wind.

* * *

Almost all the way down the eastern bank, from the Banjo pier to the bridge linking East and West Looes, the riverside is packed out with people, from toddlers upwards, each dangling a thin plastic line from a wrap-round plastic line holder over the water's edge. Each little gathering has a plastic bucket to contain their catch – crab.

A fisherman, with a mobile phone clamped to his ear, is listing his catch to a customer, possibly a local restaurant: '. . . got some doggie, little bit of mackerel . . .'. A few people are rounding up small children and dogs, packing away their gear, to return to their lodgings for tea. Just around the headland south of West Looe and Hannafore Point, Looe Island juts out of the water. According to local legend, Joseph of Arimathea landed with the baby Jesus on this island on their way to Glastonbury.

At Looe railway station, a handful of us are sheltering from the strong evening sunshine in a covered seating area, waiting for the 17.58 train to Liskeard. The Looe Valley Line has shown remarkable staying power since it first opened to freight traffic on 27 December 1860, replacing the old canal for conveying lime. Lime was needed to neutralise the acidic soils of the region for agricultural production, particularly cider-making. Officially, passengers were not allowed to travel on the trains until September 1879; in practice, tickets were issued before that date for the transport of personal belongings – hats, umbrellas and parcels – and a blind eye turned to the presence in the carriages of their owners, who travelled free of charge. The new railway was acquired from an independent company by the GWR only in 1909, by which time it had also been used for many years to transport copper mined north of Liskeard for shipping on from Looe.

The outbreak of the Second World War scotched the first plan to close this valley railway and replace it with a new rail link – by then already under construction – between Looe and Trerulefoot, near St Germans on the main line. It escaped the axe again, by the skin of its teeth, under the programme of Beeching-inspired closures, just two weeks before the scheduled closing date of September 1966.

Our train hugs the East Looe river through a heavily wooded landscape, running past the remains of the former Liskeard–Looe Union Canal – whose owners built the railway – and the museum of fairground organs, player pianos and orchestrons at St Keyne ('Paul Corin's Magnificent Music Machines'), and various other request stops. At Coombe junction, the driver and conductor change ends, and the train then reverses, passing underneath the main line into Liskeard.

The town has two stations – the terminus for the Looe Valley Line, its Victorian cast ironwork beautifully painted and the platform spotlessly maintained, and just round the corner, a few hundred yards uphill, a brand new main-line station building. Inside the latter is a cool spacious area of stylish chrome chairs and tables overlooking the cutting and platforms through a vast plate-glass window.

Although Cornwall's main-line connection with the rest of England came rather late in the day, the county had by 1850 several small independently operated minerals railways, built mainly to carry tin and copper from mines to the ports and return with coal to power the steam pumping engines. The mining of china clay grew into a major industry only in the latter half of the nineteenth century, and remains an important sector of the Cornish economy albeit on a smaller scale.

* * *

From Lostwithiel (the old Cornish name Lostgwydeyel means 'the place at the tail end of the forest' or 'lost in the hills', depending on who you believe), about 6 miles west of Liskeard, I had planned to head straight down the Fowey river and then wind my way round to Par on the coastal path. But then I'd have missed a ride on the Bodmin and Wenford Steam Railway, and the chance to experience the Camel Trail – the footpath and cycleway linking Bodmin and Padstow.

There's just time to squeeze in a detour northwards, but first a scout around Lostwithiel, an attractive little place which served as the county's capital in the thirteenth century and one of Cornwall's medieval stannary towns, the others being Liskeard, Bodmin, Truro and Helston. The term derives from the Latin for tin, *stannum.* Only these towns were authorised to weigh and stamp the tin before it went for export.

First granted its borough charter in 1189, the port by the early thirteenth century, had become the second busiest on the south coast. Edmund, the Earl of Cornwall rebuilt Restormel Castle a mile up river for his residence, and had the Shire Hall, later known as the Duchy Palace, the bridge and the church tower erected. The Stannary Parliament, made up of twenty-four representatives drawn from the Cornish tin mining districts, met at the Palace until it was wound up in 1752.

Now Edmund was clearly an astute and far-sighted town planner: what he proudly described as the 'Fairest of Small Cities' was laid out in a grid pattern, making this ancient place now surprisingly easy to navigate on foot. Throwing sticks for the dog into the river near the bridge – barely more than few inches deep at present – it seems astonishing to think that sea-going ships once sailed in and out of port here.

As a result of tin-mining activity, the river gradually became silted up and hence the port moved to Fowey. By the nineteenth century, the town was undergoing another boom, this time from the mining of iron ore, transported through the town by horse-drawn trams to the jetty. When the GWR arrived in 1859, a railway maintenance works, designed by Brunel, soon followed, providing a major source of employment to Lostwithiel men and their families for almost a century.

We head up towards the town's railway station and cross over the Tudor river bridge. While lacking any pavement, the V-shaped recesses along the bridge enable pedestrians to dodge the traffic safely. Alarmed by the sudden clunking noises from the operation of the level crossing, the dog lurches to our right, taking me with her, towards the entrance of a rather interesting-looking modern estate. A sign here says: 'From a proud past – Brunel Quays. Creating a timeless future in the ancient town of Lostwithiel. Light industrial storage units available for rent. Wombwell Homes.' The stone-clad new homes, facing the railway line and backing on to the river, seem to blend in well enough with the original railway engineering depot at the end of the road.

Coulson Park, a pleasant recreation area on the town side of the river a little way west of Brunel Quays, honours the memory of Dr Nathaniel Thomas Coulson, born in Penzance in 1853. After his mother's death and his father's desertion, the young Nathaniel was taken to Bodmin workhouse. At the age of eleven he was apprenticed to Mr Hoare of Penquite Farm, Lostwithiel. At the town's Sunday school, his teacher Abigail Santo befriended him, and he never forgot her. After serving in the Royal Navy for a few years, he left for the United States, arriving penniless in San Francisco in 1877. Having trained and qualified as a dentist he then went on to accumulate a great fortune from investment in real estate. At the turn of the century he returned to Lostwithiel bearing gifts to the town including the funding for the creation of a park. Although he lost everything in the San Francisco earthquake in 1906, he managed to prosper again and died a wealthy man in 1945, aged ninety-two. The Mayor of Lostwithiel opened the park in 1907, and the story of Nathaniel Coulson's involvement with the town was inscribed on a stone erected here, on the initiative of the town in 1998.

Fortified by lunch in the garden of the Royal Oak pub in Duke Street, we head up the lane towards Restormel Castle, Lanhydrock Woods and Bodmin Parkway station.

A mile or so along the lane, I catch sight of an elderly gentleman in a wheelchair, gazing out across the sunlit Fowey river valley and the railway line towards the golf course on the hill next to Polscoe Wood.

Quite a few golfers are striding about purposefully on the fairways, several using electric buggies between shots.

'You've chosen a good vantage point there – what a lovely view!' I say to the old chap.

'Yes, and I just seen a couple of pheasants squawking about in the field just the other side of the hedge here.' He points out the bungalow where he lives on the other side of the valley on the eastern outskirts of town – alone now as his wife recently had to move into a nursing home. When I tell him I'm walking from Paddington to Penzance, tracking the route of the GWR, his face breaks into a broad smile.

'I worked on the railways for forty-seven years!'

'No, really? A lifetime's service. Where did you work?'

'Plymouth initially, on the footplate. Wonderful years. Found it harder as time went on, all the weekends and night work – my wife got fed up with it. So we moved down here – and I got a job on the permanent way. Track work, lower paid, though we didn't mind that too much.'

We fall silent for a few moments to consider the view again, he scratches the dog under her chin and I think about moving on. Suddenly he looks up and fixes me with his rheumy blue eyes. 'Do you know? I did something this morning that I've not done for sixty years. Went to church.'

'What made you decide to go again?'

He glances up at a clear blue sky, then straight back at me: 'Time's moving on.'

Seeing what the Luftwaffe had done to Plymouth and people sleeping rough in the streets having lost their homes in the raids had cost him his faith. How any God could allow such dreadful suffering was beyond his comprehension. He seems unsure whether he'll be going back to church next Sunday. 'They do all this chanting now. It's all changed.'

'Well, you'll always have this most lovely view to come back to whatever you decide,' I say, rather lamely. Wishing him good health, the dog and I leave him to his thoughts.

The walk on to Bodmin Parkway (originally called Bodmin Road) is well shaded by mixed woodland, including sycamore and beech, and clumps of blue hydrangea blow about in the breeze along the path sides. You enter Lanhydrock Woods via a path across pastureland, leading from the end of a lane at the Lostwithiel Water Treatment Works. Right across Devon and Cornwall I've noticed that what I would call a sewage farm or sewage treatment works, tends to be called a water treatment works. Is there a subtle rebranding campaign going on somewhere to airbrush the word 'sewage' out of the English language, I wonder?

The friendly couple running the café this afternoon at Bodmin Parkway do so as volunteers, like most of those involved with the Bodmin and Wenford Railway.

While I drink my tea at the table on the platform, the husband disappears inside the former signal-box to fetch a slice of beef for the dog. Although the café is meant to shut at 5 p.m., they're sometimes still there at 7 p.m. finishing off, as they don't like to turn people away.

The Bodmin and Wadebridge was among the first railways ever built, constructed for £35,000 after a study commissioned by Sir William Molesworth of Pencarrow in 1831. The line from Wadeford to Wenfordbridge, with a branch to Bodmin, was designed to carry sand from the Camel estuary to inland farms, for spreading on the fields as fertiliser. It opened in 1834, the first railway company to be authorised for steam locomotive haulage and one of the first to carry passengers besides freight. The London and South Western Railway bought it in 1846 but it took more than fifty years for the railway to connect to its parent company's main-line operation. In the meantime, the GWR had got in on the act by building a line from Bodmin Road station into the town centre, after it eventually dawned on local worthies that the railways were no flash in the plan and, like it or not, were here to stay.

Nowadays, the preserved Bodmin and Wenford Railway is a popular tourist attraction, with steam locomotives carrying visitors between Bodmin Parkway – the out-of-town southern terminus – Bodmin General, where the main station building stands, and Boscarne Junction. From Boscarne Junction you can walk or cycle on the Camel Trail to Wadebridge and Padstow. Inspiration strikes as I walk past the cycle hire hut in the car park at Bodmin General. Recalling my delight at seeing an elderly chap cycling along the Exeter canal path near the Turf Hotel towing his dog in a trailer, I poke my head around the door of the hut. 'Do you have any trailers for dogs?' I ask the young man.

'We've got a child's trailer – would you like to try your dog for size?'

The young man, called James, fetches a bright yellow and blue child's trailer. The dog hops into it, curls up against the plastic tent-like material and looks up expectantly. What a trouper.

'Is it available tomorrow – oh, and a bike to go with it?' In a trice, James has fixed me up with a suitable bike, hooked up the trailer and lent it against the side of the hut ready for me to collect in the morning.

* * *

As we'll be catching the 11.15 a.m. steam train to Boscarne, to join the Camel Trail, there's time for a short stroll around Bodmin beforehand. The Shire Hall Visitor Centre has a huge and prominent wall display about crime and punishment in the area, in particular a rather lurid account of the popularity of public hangings that took place at Bodmin Jail. Apparently, members of the public were permitted to pull the leg of a hanging prisoner who was taking a particularly long time to die, which is how the phrase 'you're pulling my leg' came into use.

Over the road, the lady at Bodmin Museum at Mount Folly tells me that the local railway was one of the first to run excursion trips. An early 'special' train was laid on in 1840 to allow people from outlying settlements to attend the hangings at Bodmin Jail of the Lightfoot brothers after their conviction for murder. Though long closed to inmates, Bodmin Jail continues to draw the crowds to 'experience the life of an eighteenth-century prisoner, visit the gallows, stocks and underground cells, all set within three acres of original buildings and grounds', as the publicity material puts it. There appears to be no escaping tales of grisly ends, suffering and torment in Bodmin.

* * *

It's a relief to flop down on a seat in the guard's van, the dog at my feet, the bike and trailer stowed in front of us, and banish all these images of death and executions from my mind. Well, it would have been had my eyes not fallen on the blurb in the steam railway leaflet, accompanied by a photograph of a corpse on the floor, blood pouring from his head, a wine bottle next to the body: 'MURDER ON THE BODMIN STEAM RAILWAY – SOLVE A MURDER MYSTERY WHILST ENJOYING A STEAM RAILWAY JOURNEY ALONG 13 MILES OF TRACK THROUGH THE BEAUTIFUL CORNISH COUNTRYSIDE. CORNISH PASTY SUPPER SERVED AT YOUR SEAT . . . VIEW THE MURDER SCENE – CHOOSE YOUR SUSPECT.'

Really, this local obsession with gruesome murders is getting beyond the pale. Obviously the poor fellow in the photo has been battered to death by a bottle-wielding madman. Or has he? Perhaps he's just pretending to be dead, having just bumped off someone else. I find myself glancing at the upcoming dates for murder mystery pasty supper specials. Oh – I see they run murder mystery cream tea specials too. Enough.

Having taken the precaution of lining the trailer with the dog's fleecy mat, I'm expecting maximum cooperation from the animal after alighting to join the cycle path at Boscarne. She steps into it obediently, humouring me I expect, and off we go. After conveying the dog in her chauffeur-pedalled chariot for a minute or two, I sense a commotion stirs the back. I turn round to see a wet black snout, a pair of brown eyes and two grey front paws poking through the Velcro fastening of the trailer flap. Sensing mutiny in the ranks, I dismount, open the flap and let Fly jump out to run alongside the bike for a while, which she seems to prefer. The only hitch is when other bikes approach and she dives for cover in the bushes, which slows us down somewhat, as she is then reluctant to come out again.

We attempt a compromise. She goes back on the lead, runs alongside the bike, apparently content, leaving me just one hand to steer the bike. (I later discover that this behaviour breaches the Camel Trail Code of Conduct: 'Cycling with your dog on or off a lead is very dangerous,' it says.) By the time we arrive in Wadebridge, 5 miles down river, Fly is ready and willing to hop back into her chariot, curl up, admire the scenery like Lady Muck and leave me to do all the hard work.

Just as I'm looking about me to push the bike across the road towards the library, I catch sight of what looks like an old railway station down Southern Way, an otherwise unremarkable residential road to my left. Inside we are greeted by some charming ladies of some maturity, one of whom behind the reception desk asks if I would like tea or coffee – 30p including biscuits. Or I can have soup and a slice of toast. It being lunchtime, I opt for the soup, slice of toast and the cup of tea with biscuits over in the sun lounge where there are magazines and newspapers spread over some tables for visitors to browse. What a great place.

So complete is the conversion, that were you to be guided into the building blindfold, you'd never guess that the building once accommodated both the GWR and the Southern Railways. After the last passenger train left Wadebridge station late afternoon on 28 January 1967, the building fell into dereliction for several years. After four years of fund-raising, spearheaded by a retired GP, Dr Gordon Kinsman-Barker, and his wife Mary, the building was acquired by the charity they helped to form, Wadebridge Concern for the Aged. The charity converted the station for use as a day centre for recreational and educational activity, and it reopened in 1989. One of its rooms houses the Sir John Betjeman Centre, a collection of personal memorabilia – mementos, furniture and academic honours accumulated by the late Poet Laureate, who lived in the area towards the end of his life. There's a rather formal-looking photograph on the wall of Betjeman and Philip Larkin in university gowns, having each accepted honours from some institution or other; can I detect a suppressed guffaw in their politely fixed expressions to camera?

Under the auspices of the University of the Third Age, the centre offers around thirty classes a week ranging from embroidery to keep fit and there's a tea dance on Fridays. For £15 a year, you can attend any class or all of them, and some of the students are well into their eighties. Nine volunteers, operating on two-hourly shifts, run the centre each day. There's a large photograph on the wall in reception of the then Prime Minister John Major's visit in 1993.

Hilary Mulley has been helping out in the kitchen for four years and attends classes most days: 'It's a marvellous centre, and a great new use for the old railway station. People enjoy dropping in, seeing their friends and having a chat. We're very lucky to have it.'

Dora Ellis, now in her second year as a volunteer, teaches her students how to make their own greetings cards. 'I wanted to do voluntary work here when I was still working, but it's only been since I retired that I've had the time and opportunity. It's a great place to meet people and get to know the town, especially when you've moved down here from up country, as I did a few years ago.'

The onward ride to Padstow, apparently the most scenic section of the trail, will have to wait until another day as we're running out of time. The last steam train from Boscarne back to Bodmin General is just about to leave, which means we've missed it. A longer return bike ride along the trail to Bodmin town centre therefore lies ahead. This is no hardship as the path is excellent, the woods of the river valley providing welcome shade on another hot day. They've thought of everything here – even mounting blocks for horse-riders near junctions with roads and other tracks.

The dog's as good as gold on the way back, slumbering peacefully all the way in her chauffeur-pedalled chariot. Just as well, since pushing her and the bike around Bodmin during the rush hour up the steep incline of St Nicholas Street, back to the cycle hire hut, is hard enough as it is.

The only troubling thought as I pause to mop my brow up the endless hill to Bodmin General is that, on presenting myself at reception, how swiftly the mature ladies at the Wadebridge Concern for the Aged day centre had offered me subsidised refreshment. They must have taken me for a citizen of the Third Age. Me – a mere forty-something, still in the bloom of middle youth. What a flaming cheek!

10

Exploring the Copper Kingdom
Lostwithiel to Penzance

In which we discover an awesome lunar landscape north of St Austell; skirt the foot of Carn Brea in swirling mists; admire a statue of the 'Cornish giant' and give a student from Vancouver Island a garbled introduction to Brunel and the GWR.

Enjoying her middle youth to the full, Fly scampers about the path through Coulson Park, Lostwithiel, rounding me up from the rear, lying down, ears flat, when she's found a stick, busy working. This is my cue to *fetch* the stick, pick it up from the ground – or pluck it from her mouth – and throw it for her. This game only ends when the stick retriever – ultimately me – is too weary to carry on and begs for the dog's mercy. A couple of young lads are hurling much larger sticks into the lower branches of a horse chestnut and then jumping up and down on the fallen fruit to flesh out the conkers.

Flunking the temptation to ask them whether they should not be at school, I call Fly to heel and together we bear right through a pedestrian tunnel to reach the public right of way leading leftwards towards Par, a very pleasant up-over-and-down stroll through largely deserted farmland, sunken narrow lanes and canopies of hawthorn, blackthorn and bramble, as it turns out. Initially we walk through woodland between the old mineral railway and the GWR main line, the River Fowey and marshland south of Lostwithiel on our left. There's no one about, we're enjoying what's become a reassuringly familiar combination of background sounds – birdsong interspersed by occasional passing trains.

Along a wooded path, we cross several small streams flowing across large flat stones. By the time we emerge into pastureland, the sun has burnt off the mist and drizzle of the early morning and lit up the river valley. On the other side of the Fowey lie marshes and fields rising steeply to meet blue skies and billowing white clouds. Back on the river bank, under the shade of the trees, we find a level path hewn out of the hillside, covered in leaf litter, and soon the massive stone arches of the Milltown railway viaduct dwarf us.

We reach a junction of narrow minor roads and turn left to join the Saints' Way. Originally a Bronze Age trading route linking the north and south Cornish coasts from Padstow, via the Camel and Fowey rivers, it was later used by Celtic missionaries preaching Christianity to the heathen. The rough granite crosses, standing stones and holy wells along the route, indeed right across the county, signify their enduring influence. A fairly strenuous ascent to the brow of the next hill is rewarded with magnificent views down to the river – now much wider than the modest affair running through Lostwithiel. Our footsteps drive several young game birds out of their cover in the verges and hedges, squawking and flapping furiously.

Having joined Passage Lane on the northern edge of Fowey, I soon glimpse some china clay lorries rolling along in near silence up and down a road beneath me to my left, laid on the route of an old railway into town. The former stationmaster's house at Fowey is boarded up and the garden heavily overgrown. A notice warning the public to keep out is attached to the gate. A car park and public library occupy much of the former station site, although the only connection between the nearby Chuffers restaurant and the old station is one of proximity.

A monster traffic jam is building up nicely in Fore Street, as drivers wait for vacant spaces in the car park and holidaymakers weave a path through the stationary cars. The town quay is full of people scoffing fish and chips and shooing away the predatory seagulls that seem to double in size every time I visit Cornwall. The dog wolfs down the thick semicircular crust of my 'traditional' Cornish pasty while I polish off the steak, swede, onion and potato filling.

In and around Fowey, the transport and export of china clay appear to operate at a discreet distance from the historic heart of this beautiful town of narrow streets and alleyways. Unless you stray off the beaten track – and few visitors appear to do so, at least on today's evidence – you'd barely know it was there. At Par, just around Gribben Head and its most visible landmark – the fine red and white day marker – the industry's vast operations dominate the land and seascape. A private road running east–west (cross it at your own risk, says the sign), and used exclusively by the china clay lorries, annexes Par beach and the 1-mile long stretch of coastal land accommodating a caravan and camping park – all private territory.

The dog and I cross the private road, at our own risk, to see if there is a way of rejoining the coastal path without having to trudge along the A3082. Sadly, the unofficial tree-lined paths that tantalise explorers of the southern side of the private road lead only to uncrossable stretches of water. So it's back to the busy main road, under the railway line, twice, and past the entrance to Par harbour, built in the early 1820s by Joseph Thomas Treffry to service his copper mines and granite quarries.

Back on the coastal path, the dog kicks her heels, lopes away from the main road, sniffs the bottom of a hedge and gives me a toothy grin. We pass the south-western reaches of the vast grey Imerys works, and the eastern fairways of the Carlyon Golf Club on a section of the path that conjures up images of a frosty January morning. Every leaf and blade of grass within sight is coated with a layer of fine white deposit.

We emerge on to a gloriously empty, almost abandoned, stretch of beach around Spit Point. The tide is going out, and a pair of egrets are quarrelling over a morsel of food close to the water's edge. To the north I can see rows of caravans at the Mount Holiday Park. Roughly halfway between Polkerris and Gribben Head, to the south-east, stands Menabilly, once the home of Daphne du Maurier and the inspiration for Manderley in what is perhaps her best-known novel, *Rebecca*.

On the golf course we meet an elderly lady and her terrier Minnie, whose passion in life is to retrieve lost golf balls from the bushes on the cliff edge. They live in Par, having moved there recently from Charlestown, which she found too crowded in the summer. Through a gap in the bushes, we peer down at a rather sad-looking stretch of beach, dominated by a large, ugly and derelict building that seems as though it could have been wrecked by some sort of catastrophic event. What a depressing shambles it appears – the beach enclosed by bits of temporary metal sea defences, and a metalled service road curving around the foot of the cliff.

'Is this an abandoned outpost of the china clay works or something?' I ask Minnie's owner. She sighs and shakes her head. Carlyon Beach was a popular family resort in the 1930s, and the Cornwall Coliseum an entertainment venue, attracting, she recalls, stars like Tom Jones and Shirley Bassey. For many years a firm of developers, Ampersand, has been planning to build more than 500 apartments here, serviced by shops and new leisure facilities. The proposed £200 million development called 'The Beach' is currently on hold while the Government Office of the South West decides whether ministers should determine Ampersand's associated planning application for improved sea defences. Minnie's owner tells me she thinks it outrageous that a private holiday village could be built on what used to be a public beach, but expects it will go ahead anyway.

On the outskirts of Charlestown, I come across a sign pointing towards a lookout tower a little way to the left of the coastal path: 'National Coastwatch Institution – Visitors welcome'. The dog and I climb the steep metal steps to be greeted by a friendly volunteer watchkeeper. We catch Michael Down, a retired nurse, halfway through a 4-hour stint surveying St Austell Bay in a little station equipped with telescope, phone, radar, weather-recording instruments and shipping charts to plot the course of visiting craft. Now here's a familiar story of a once-proud public sector service, funded by the taxpayer, cut to the bone in the name of rationalisation and efficiency, leaving unpaid volunteers to pick up the pieces.

The NCI, a voluntary organisation, was formed in 1994 after two fishermen drowned off the Cornish coast, within sight of an abandoned coastguard station, closed as a result of budget cuts. John Major's Conservative government, ever anxious to follow Lady

Thatcher's raids on public-sector budgets at every opportunity, had decided to shut down several of the smaller coastguard stations.

Michael Down is one of twenty-two trained volunteers operating the Charlestown NCI station, part of a national force of more than 1,000 coast-watchers.

'We look out for any incidents in the bay, anyone in distress, keep an eye on the boats,' he says, inviting me to have a look through the telescope. There's always some activity, naval training exercises, submarine activity, fishing, swimming and so on to engage the coastwatchers' attention. 'Anything that moves basically, we monitor, record and where appropriate report to the coastguard at Brixham. Any suspicious activity that might interest Customs and Excise, we can alert them directly. We get quite a few visitors to chat to, so no, I don't get lonely. Ours was one of the first NCI stations to open – there's another at Polruan, the other side of Fowey, one at Penzance. It's very enjoyable work.'

After signing the visitors' book, underneath an appreciative comment from an Air Commodore, we take our leave. Michael encourages me to fix my gaze on the road ahead as I lurch very slowly down the steps, clutching the handrails and trying very hard not to look down. He follows behind keeping the dog on a short lead. At Charlestown harbour, built by Charles Rashleigh for the export of china clay, two tall ships are open for visits, the *Kaskelot* of Bristol and the *Earl of Pembroke*, as is the Shipwreck, Rescue and Heritage Centre. I decide to make another detour and go in search of a bus stop. It's time I found out more about china clay.

* * *

Wheal Martyn China Clay Museum at Carthew, on the western banks of St Austell river, has been extended and rebranded since my last visit a few years ago. It's got a new circular white entrance building, accommodating an exhibitions space and toilet block, all very spacious, light and airy. By the entrance is an arresting piece of art depicting a group of Victorian china clay workers. The fibreglass models were dressed in real clothes, and then the whole ensemble sprayed in liquid china clay. Eerily lifelike, as well as life-sized, these clay figures have on occasion scared the living daylights out of a lone member of staff or visitor arriving at night.

Now, in addition to 'sewage', here's another word that appears to be losing favour in many places: museum. This one now styles itself as the China Clay Country Park, Mining and Heritage Centre. If that helps to project a more exciting image of what's on offer, and gets more visitors of all ages through the door, well, who am I to object? Whatever one chooses to call it, this place is a gem, a real pleasure to explore, inside and out, just as I remember it only better. It makes learning fun and, moreover, it's dog friendly.

Used today in the manufacture of pharmaceuticals, cosmetics, paper and newsprint, vacuum cleaners, uPVC, road markings, tableware and sanitaryware, china clay – or *kaolin* – is a powdery white mineral formed by decomposing feldspar, a constituent of granite.

In Cornwall it was originally used to line smelting furnaces, as a by-product of quarrying china stone for building material. Other potential uses were first identified by William Cookworthy, a Quaker apothecary from Plymouth, after visiting some mining works near Helston in 1746, and others later around St Austell. As a result of his experiments, he created the country's first hard paste porcelain, patented it and opened his own pottery at Coxside in Plymouth. More uses were discovered, particularly for making paper and cotton, and by 1858 some forty-two companies had joined the 'clay-rush', producing 65,000 tons a year of this valuable commodity. After Cookworthy retired, his business partner Richard Champion was foiled in his attempt to extend the patent following opposition from the Staffordshire potters, led by Josiah Wedgwood.

Now, with fourteen working pits, Imerys is the UK's largest china clay producer, using a fleet of sixty-five vehicles. In terms of getting the finished product to customers, domestic and worldwide, it's mostly done by sea and rail.

Beforehand, however, there are numerous stages of processing, refining and conveyance along 250 miles of pipelines, and 350 miles of private roads, linking pits to works. The best way, I find, of trying to get your head around all this is to step out on to the nature trail in the direction of Pit View, ideally with binoculars at hand. In 1955, some 1 million tonnes of china clay were produced a year – now output stands at an annual 2.5 million tons. The Cornish clay pits are said to be good for another forty years of working. Thanks to modern filtering systems, no longer is the St Austell known as the 'white river', although, as I've seen, particles of dust still deposit themselves on leaves and blades of grass, presumably elsewhere too.

After viewing the exhibits in the interactive visitor centre, basically the indoor museum, you then find yourself ready to explore the rest of the complex. First you see a monitor jet, a rather menacing-looking python-like coiled tube used to shoot water at high pressure at the base of the pit. This procedure creates the clay slurry, from which the useful mineral is extracted.

On reaching Pit View, an extraordinarily vast lunar-like landscape opens up, the old workings colonised by plant life, green patches turning to grey, and current activity distinguished by off-white, sharply angled layers of exposed mineral. A monitor jet is squirting water at the rock in Greensplat pit on the left, remotely controlled by an operator sitting in a nearby hut. I can't remember seeing anything remotely like this before, but the scene conjures up childhood memories of fuzzy TV images of the first Moon landing and later Apollo missions, and of spacecraft crawling and tottering about the lunar surface, scooping up tiny samples of mineral deposit. And yet both Greensplat and, on the right, Wheal Martyn pits are small compared with the 'superpits' of Littlejohns and Blackpool a short distance away, apparently.

The dog and I are staying with my cousin Roger Meldon-Smith and his wife Liz, and their spaniel puppy Tara. Their house, halfway up Bodrugan Hill at Portmellon, near Mevagissey, overlooks both the cove and the village. In the evening, we visit an old friend of theirs staying at a nearby holiday cottage by the seafront and sit outside chatting and soaking up the bay view until darkness descends. Liz describes what we can see on the horizon.

'See those lights on that hill, that's Polkerris. And the pink glow in the distance, that's Plymouth. How was the China Clay Museum?'

'Extraordinary. Plenty to see and do, inside and out. Great café.'

* * *

Some of the V-shaped stone stiles along the footpath between Portmellon and Gorran Churchtown look as though they may have been there since time began, the surrounding hedgerow beautifully maintained and clipped – rural topiary at its finest. Bribed with much stick-throwing activity and coercive babble from me, the dog more or less obliges my attempts to photograph her leaping and scrambling over these stiles. Along this stretch appears the most unequivocally hospitable message to walkers that I've seen since leaving Paddington station: 'Woods on your Doorstep. The Woodlands Trust. You are welcome to walk here any time.' The contrast with some of the finger-wagging 'Keep out' notices posted along certain stretches of the Home Counties, couldn't be more striking. Sadly, the prospects for walkers onward bound from Gorran Churchtown to Truro take an immediate turn for the worse. Almost every possibility is by minor road, sunken between high Cornish hedges, offering few views and landmarks to enjoy or use for orientation. Unless you have limitless time at your disposal, in which case the coastal path and King Harry's Ferry might be better alternative routes, it's probably best done by bike.

* * *

Those early Christian missionaries, and their devotees, knew a thing or two. The relentless peal of church bells calling people to worship could wear down the resistance of the most determined pagan. Desperate to flee the deafening clang of Truro Cathedral bells, the dog pulls me on the lead away from the city centre. We pass the old County Hall building, fringed with palm trees, and soon the new county administration offices, to join a footpath at Penwethers. We cross the main line west of the viaduct near Newbridge, and zigzag our way towards Redruth through a gentle green patchwork of sunken lanes, bridleways and farmland.

Soon we enter a wilder-looking landscape of bracken and gorse, ruined mine stacks and engine houses, rising above a granite moorland criss-crossed by a tantalising array of paths. Directly in front of us is a granite marker post with a black engine-house logo carved on the stone: we've reached a junction with the Mineral Tramways Coast-to-Coast Trail. A couple on bikes freewheel down a gentle incline from the direction of Scorrier, wave as they pass us and disappear round a bend in the trail path towards Devoran. We aim for Redruth.

The town acquired its Cornish name from the discolouration of the local brook by the release of iron oxide from tin streaming – *rhyd* = ford, *ruth* = red. The advent of deep mining for copper from the 1730s transformed the economy of Redruth and nearby villages, as it did around the brass mills between Bristol and Saltford. For more than a century, the vast tracts of inhospitable moorland surrounding the town were dominated by mines, specifically the Gwennap mines, and the heart of Britain's 'Copper Kingdom'. What enabled this rapid industrialisation of the landscape to take place were the railways – and their earliest incarnations, the tramways. By the early years of the nineteenth century, packhorses plodding along rough tracks could no longer cope with the huge volumes of ore that needed to be shifted to the coastal ports for despatch to the smelting works in South Wales.

Cornwall's first mineral tramway, modelled on a system already operating in South Wales, was built between Portreath harbour on the north Cornish coast and Poldice Mine 4 miles south-east of the port. The first rail was laid in 1809 and the whole line in use by horse-drawn wagons by 1819. Between 1801 and 1866, the number of Cornish mines mushroomed from 75 to 340, and they produced two-thirds of the world's copper when output peaked towards the end of this period.

The typical miner endured dirty and dangerous working conditions several hundred feet underground. Having been sent down the mines at the age of eight, he was unlikely to live beyond the age of forty. When the industry began to decline in the 1860s, amid growing international competition, many mining families in and around St Day emigrated to Australia, South Africa and North America, giving rise to the old saying: 'Wherever in the world there's a hole in the ground, you'll find a Cornishman at the bottom.'

The Portreath tramway operated for about sixty years. This freight-only line was horse-drawn at first but in 1854 converted for use by steam locomotives – *Smelter, Miner* and *Spitfire* – until its closure in 1915.

Today's footpath and cycleway continues from Poldice down to Devoran on the Restronguet Creek along the line of another former mineral railway, the Redruth and Chasewater. Heads down into the wind and drizzle, beneath threatening overcast skies, Fly and I cross over the former mineral tramway to follow a path uphill, past an old mineshaft covered by a flattish cone-shaped metal cap, in the direction of St Day, soon to enter the eastern reaches of the village via Higher Goongumpas Lane.

The problem I have here is that no matter how many times I turn the map round and trace various green dashed lines with my finger, I can't see a way of getting to Redruth that doesn't involve walking on several roads, possibly in darkness. The driver of a parked Royal Mail van advises me not to attempt it. 'All these minor roads round here are used as rat runs at this time of day, and some people drive at horrendous speeds. I don't think you'd be safe, especially walking with a dog.'

'Any chance of a lift then?' I ask, with a hopeful smile.

'Sorry, this is a government vehicle, we're not allowed to take passengers. Look, there's a bus shelter over there.'

I scour every vertical surface of the bus shelter for information about what buses stop there, where they go and at what time, and find none of any description. With traffic streaming up and down the B road from Higher Goongumpas Lane into the centre of St Day, and no sign of pavements on either side, it takes me a while to work out a circuitous route, past a caravan site.

Eventually, we find ourselves at the entrance of the splendid St Day Church, where there's a description of the area's history on a board, and another bus stop, one with a timetable displayed. Just as I am beginning to doubt that it will ever come, and having failed to raise First Great Western customer services on the advertised number, the number 40 bus to Redruth and Camborne trundles into view 15 minutes late, packed with shoppers and schoolchildren on their way home. The dog drapes herself across the feet of two seated passengers, one of whom says: 'No, it's all right – she's keeping us nice and warm!' The bus deposits us at Redruth railway station near the viaduct.

The completion of Brunel's original wooden viaduct in 1852 enabled passenger services to Penzance to begin in March of that year, and the timber structure remained in use for thirty years. Its replacement, a towering edifice of local granite with 70ft-high arches straddling streets of terraced houses, took four years to erect.

* * *

Margaret Johnson's directions from Redruth railway station to her B&B in Trevingey Road are excellent. You go downhill to the traffic lights and then left underneath the railway viaduct, first right past the Cut & Dried hairdresser's, on to a woodland path up to Blight's fish and chip shop. Having spent many holidays in Cornwall, Margaret and her husband Graham moved down here from Preston, Lancashire after he retired from the police force. They invite me to join them for dinner, during which it emerges that Margaret knows my cousin Roger's wife Liz, as they were both students over the past year on a theological course.

The dog has been fed, watered and quartered for the night in the Johnsons' garage. To my shame, she blew her chance of sharing the utility room with the couple's elderly and very sweet labrador Bruno, after sealing their acquaintance with an entirely unprovoked snarling charge at him. Bad girl. In the morning Graham describes to me the best walking route to Camborne, around the foot of Carn Brea, and Margaret tells me to watch out for her church en route. 'St Euny's was the original parish church of Redruth, and there's been a church on that site since AD 600,' she says. 'The only reason it stopped being used as the parish church is that the mine owners had their houses built along Clinton Road, and so they then had St Andrew's Church built, which is like a mini-cathedral. But ours is the mother church, if you will.'

Fuelled by an excellent breakfast, I step out on to the Trevingey Road towards Carn Brea village. Fly, forgiven for her misdemeanours of the previous evening, pauses to drink from a spring on our left – the holy well of St Euny, a Celtic missionary who visited the area in AD 500.

Disappointingly, most of Carn Brea, which rises to 738ft, is shrouded in swirling mist, but at least the rain is holding off. Excavations several years ago revealed an extensive walled Neolithic settlement, the earliest of its kind in England. The path we're following, running alongside the northern edge of the hill, is lined with massive chunks of granite. Large cobwebs, still glistening with the morning's dew, are stretched like muslin across clumps of brambles, gorse and heather, flickering in the breeze. A short flight of old granite steps, and a section of kerbstone, rise and then vanish into the moorland on my left. Further on a vast industrial area opens up to my right, where I see some kids on top of a spoil heap staring at me.

On reaching the Pool side of Camborne, a sign tells me that I've just walked along part of the Great Flat Lode Trail, a 6½-mile long circular route around Carn Brea hill. The Mineral Tramways Project has opened up more than 30 miles of old tramways and railways like this and one from Portreath to Poldice Mine, enabling walkers and cyclists to discover this extraordinary long-abandoned world of ruined chimney stacks and engine houses around Camborne and Redruth.

My attempts to find the Mineral Tramways Centre, near the railway mainline, come to nothing. It appears to have been relocated to the Mill and Mining Museum at the old King Edward Mine, Troon to the south of Camborne. The recorded message I hear on my phone informs me that this museum is closed today, so we climb the footbridge over the railway, and the remains of Carn Brea station, into Pool and see what we can find there.

The more you explore inland Cornwall, the better you can appreciate its talent for nurturing innovators, engineers and entrepreneurs of all kinds: their achievements are celebrated not just in museums and heritage centres, but in plaques, statues and monuments, town trails and restored buildings all over the place. Pool lies more or less at the centre of the conurbation of Redruth and Camborne, now joined at the hip by sprawling development. The dog and I head up Carn Brea Avenue, past the Pool Leisure Centre, to check out a couple of the landmarks of the Pool Trail. A modern bungalow now stands on the site of the former safety fuse factory. William Bickford, who came from Tuckingmill in Camborne, invented and manufactured the safety fuse, which has saved many miners' lives all over the world.

Further along the avenue, the Carn Brea engineering works once produced steam locomotives for Brunel's GWR. North of the leisure centre, I notice a memorial stone marking the birthplace of Richard Trevithick (1771–1833), inventor of the high-pressure steam engine, and erected in 'appreciation of the great inventor and his gifts to the world'. He was known as the Cornish giant. A handsome statue of Trevithick, clutching a model of his steam locomotive, stands on a plinth in front of Camborne's public library. A forerunner of the motor car, his high-pressure steam locomotive, known locally as the Puffing Devil, was given its test run on Christmas Eve 1801. Three years later he tested the first passenger-carrying steam locomotive to run on rails. The engine hauled 10 tons of iron, and seven men, spread over five wagons, a distance of 9 miles. He died penniless in Dartford, Kent.

Between Pool and Camborne town centre, a grim complex of industrial estates and retail parks – B&Q, Halfords, Homebase and so on – and a proliferation of signs and hoardings partially conceal the fragile remains of old engine houses and chimney stacks. At South Crofty Mine there's a sign inviting people to take an underground tour of the workings.

Of course it wasn't just advances in railway engineering that sustained Cornish mining through the boom years of the nineteenth century. In order to meet ever-rising demand from the industrialising countries, mining engineering and technology – the machinery, tools and processes – underwent almost constant development and refinement to keep ahead of the competition. The Holman brothers of Camborne helped to keep the local mining industry in business for a further forty years after production peaked with their development of the rock drill, powered by compressed air, which sold worldwide.

The historic town centre of Camborne is quite a compact area of fine old buildings, many of them sympathetically restored, a few empty and neglected. The Centenary Wesleyan Chapel opposite the end of Trelowarren Street, and Camborne Library, are exceptionally beautiful. The railway station, south of the shopping centre, looks decidedly spartan and in need of cheering up, especially on a gloomy autumn afternoon like this one.

Halfway down Trelowarren Street, I tether the dog to a street bench and nip into a café for a Cornish pasty and cup of coffee. People seem very friendly and easy-going in Camborne, I find. Casting about the room for somewhere to sit, I spot an elderly lady having her sandwich at a table in front of me nearest to the open doorway. 'Come and sit next to me,' she says, 'Then you can keep an eye on your dog. She looks quite happy there lying under the bench.' An ambulance arrives within a few minutes of another diner across the room collapsing to the floor, hitting her head on a table on the way down. My bill comes to £2.25. The dog's patience is rewarded with offerings of torn-off bits of pasty crust.

* * *

Although it's not quite yet 4 p.m., it's almost dark and beginning to rain quite heavily. More worryingly, the dog, unusually, has her tail between her legs and looks completely fed up. The plan was to try to push on to Hayle, but with a miserable dog to deal with and the heavens opening, I need to get us to our B&B stop sooner rather than later.

We've just missed the last bus to Leedstown, but from Camborne bus station I'm directed to the taxi rank outside Argos, round the corner from Commercial Square. The dog manages to fold herself into a neat oval shape, just filling the front passenger seat's footwell of the taxi. I spread a damp OS map across my knees, brief the driver then off we go.

'A woman who can read maps!' exclaims my driver, to whom the dog, inexplicably, has taken a shine. Fly looks up at him adoringly, ears cocked, craving his attention. I lock my right hand on to her collar, struggling to stop her jumping on to his lap. 'Well, just let me know where and when to turn off.'

'Will do. By the way, with the mining gone, what does Camborne actually produce nowadays?'

'Babies. You've got fifteen-year-old girls in Camborne, with not one but two kids in tow. There are a lot of care homes too. People often move to Cornwall away from their families to retire, and after a few years maybe, the old ladies are widowed, and they've got no one to look after them at home. Compared to the coastal towns and villages, Camborne doesn't get many visitors. It's a shame in a way, because it's not a bad place really.' To find suitably dog-friendly accommodation for the night, we're having to travel 6 miles or so south of Camborne to Little Pengelly Farm near the hamlet of Trenwheal, which is just off the B3302.

Maxine Millichip and her partner Robert make us very welcome and comfortable. Having quit good jobs and moved down here from Surrey only in February, they've just spent their first season of self-employment providing B&B accommodation, self-catering cottages for holiday lets and farmhouse cream teas: their neighbour's Jersey and Guernsey cows provide the divine clotted cream.

* * *

After a good night's sleep, the dog and I are raring to get going again. The weather forecast warns of showers mixed with sunshine, but so far it's bright and dry. From the bridleway above Lelant Downs, I recognise Godrevy Lighthouse on the north-eastern horizon, waves lashing the rocks on which it stands. To the east, the summit of Carn Brea is revealed at last, as is its monument erected in 1836 to the memory of Lord de Dunstanville of Tehidy. Formerly Sir Francis

Basset, this benevolent landowner is remembered for his contribution to the welfare of local miners.

Soon after joining the St Michael's Way, my romantic image of the archetypal Cornish stile – a reliable, solid, walker-friendly marriage of craft and nature – takes a bit of a battering. The one in front of me would be more accurately described as a large pile of stones straddled by two lengths of barbed wire fencing. While the dog takes the whole thing in a couple of graceful leaps, I inch my way across the stones in an ungainly scramble to avoid being impaled. Great flourishes of deep-pink fuchsia run riot through the hedgerows, still heavy with blackberries, the lush green leaves slowly curling and fading to golden brown. On a stile near Ludgvan a small walkers' sign describes the St Michael's Way as part of a European cultural route and a 'pilgrim route to Santiago'.

The White Hart Inn at Ludgvan, which dates back to the fourteenth century, smells of beeswax polish, old furniture and cooking. Given that the dog has been rolling in something very unsavoury this morning, possibly badger poo, I am amazed that we have not been ushered to the outside tables.

As the farmland begins to slope gently down to the sea, I see a semicircle of surf washing the rocks of St Michael's Mount, then a spectacular view of the coastline from Marazion almost all the way past Penzance and Newlyn to Land's End. Two-carriage trains glide up and down the main line between the A30 and the beach.

One of the four archangels, often leading the heavenly host, St Michael is said to be the protector of Cornwall as well as the patron saint of mariners, the airborne and grocers. Since the rainclouds of the early morning have drifted elsewhere, possibly to empty over St Ives, and the sun is beginning to illuminate the coastline on the last day of my walk, I have decided to adopt St Michael as the patron saint of long-distance walkers and their dogs.

'That's brave of yer walking from London, on yer own!' exclaims an elderly chap, with a short grey beard, three excited spaniels weaving around his heels. I meet him and his friend near the railway line and the Mount View Hotel, which he recommends if I decide to stay on overnight. 'Wor yer not wurried about meeting trouble on the way? Did yer dog protect yer?'

'Do you know – almost without exception, everyone's been very friendly and helpful, or they've just ignored us. I've never felt at all unsafe. The dog's been a star.'

* * *

For the last time, we stop, look and listen before crossing the main line, and head down to the beach. Although the summer ban on dogs on the beach is not formally lifted until tomorrow, the place is overrun with them, leaping about in and out of the waves, fetching and carrying balls. Unable to find any sticks, mine presents me with disgusting bits of seaweed to hurl for her.

We continue along the beach and join a path running alongside the railway line towards Penzance station, and beyond it the familiar landmarks of out-of-town ribbon development, Tesco, B&Q, Kwik-Fit; then the fire station and another National Coastwatch Institution station looking out to sea high above the rooftops. A group of camera-wielding railway enthusiasts are watching and waiting in silence on the path for any movement on the line worth photographing.

We arrive at the station to catch the 16.00 to London Paddington, just in time to hear the rain clattering down on the roof like great volleys of ball-bearings. From the concourse, I look along the platforms, beyond the roofline. A rainbow stretches from St Michael's Mount over the line and disappears into the hills around Ludgvan.

Hanging about on platform 4, while our waiting high-speed train is cleared of litter, I ask a friendly looking young backpacker if he would mind taking our picture by the Penzance sign. Braden Hutchins, aged twenty, from Vancouver Island, charmingly obliges. He's been visiting relations around Penzance, before starting a study course on Monday on environmental management and urban design at the famous Camborne School of Mining, now just outside the town at Pool. For the rest of the weekend, he's staying with friends in Plymouth.

'This will be my first ride on a British train,' he announces, as the three of us settle down in the 'quiet' carriage, the dog arranging herself underneath his seat.

'Well, you're in for a treat,' I say, somewhat prematurely. 'Fantastic scenery. Loads of viaducts over river valleys. The Royal Albert Bridge over the Tamar. I'll be your tour guide!'

Just before the train gets going, eight rowdy, foul-mouthed, middle-aged drunks lurch past us and fill the table seats a few feet away. A harassed-looking train guard works his way down the aisle, clipping tickets.

'Don't worry,' he tells us, before moving through to the next carriage. 'They're getting off at Truro.'

Braden looks shocked at our neighbours' loutish behaviour, while I just feel wearily embarrassed. We distract each other from it in wide-ranging conversation, encompassing British and Canadian politics, the scariness of George Bush (and Tony Blair), the absence of both railways and human settlement from vast tracts of Canada, global warming, recycled maggots and the view from the window. Braden Hutchins hears a lot from me about the view from the window, probably more than he would like. The train slows as we approach the viaduct at St Germans, and once on it we peer down at the River Tiddy and the cottages along the quayside 'That's really beautiful,' he remarks politely.

'Just keep looking. I don't want you to miss Brunel's bridge. The track curves round to the right after Saltash station, so you'll see the front of the train joining the bridge. Any minute now.'

With necks craned and noses against the glass, we're like a couple of kids staring through a sweetshop window. Over the River Lynher, through the tunnel, on to Saltash, past the lane that leads to Mary Newman's Cottage, little snapshots of landscape of my long, rambling, leisurely stroll, flash before us. 'See the swans . . . the Union Jack covering the front of that pub down there? The bridge . . . They carried him over it when he was dying. He was only fifty-three . . . Did you know his mother came from Plymouth?'

Bradon frowns as he surveys the fast-changing scenes, waiting for my stream of excited babble to abate. Spinning round in my seat, I rabbit on, unable to contain myself: 'And look – there's the road bridge, built just over a century later. The road's a bit higher than the railway here – you don't really appreciate that when you're walking underneath the bridges. Do have a look round under the bridges, if you ever get the chance – it's just fantastic.'

The train rolls smoothly, almost silently, through the housing estates of St Budeaux, past Devonport dockyards and finally slowing into Plymouth station. 'Amazing,' he says finally, after I've sunk back into my seat, inadvertently waking the slumbering dog. Braden rises from his seat, swings his backpack over his shoulders, shakes it into place, and joins the line of passengers grouping along the gangway, gathering up their belongings, turning to face the carriage door, ready to disembark. We wish each other farewell and good luck.

'What was that guy's name again, the one you say made the rail bridge?'

'Brunel. Isambard Kingdom Brunel, engineer, 1806 to 1859. A great man.'

Places of Interest
& Contact Information

1. THREE BRIDGES & A STATUE

Kensal Green Cemetery
Once described as England's most important cemetery, this beautiful 77-acre Victorian memorial park is the perfect spot for a peaceful stroll either under your own steam or in a guided group. It is open daily, with guided tours on Sunday afternoons. Contact the Friends of Kensal Green Cemetery, c/o The General Cemetery Company, Harrow Road, London W10 4RA. www.kensalgreen.co.uk

Friends of Friendless Churches
St Mary Magdalene Church, Boveney, is one of the many historic places of worship saved from demolition or decay by this charity that works in partnership with the Ancient Monuments Society. Friends of Friendless Churches, St Ann's Vestrey Hall, 2 Church Entry, London EC4V 5HB. www.friendsoffriendlesschurches.org.uk

2. CUTTING A DASH ALONG THE THAMES

Cholsey and Wallingford Railway
The Cholsey and Wallingford Railway Preservation Society has revitalised a 2½-mile-long branch line and organises an annual programme of open days and special events. It hopes to build a new 'period' station at Wallingford to replace the huts. Further information from the CWRPS, PO Box 16, Wallingford, Oxon OX10 9YN. www.cholsey-wallingford-railway.com

Didcot Railway Centre
Run by the Great Western Society, pioneers of railway preservation, the centre has created a living museum of the GWR, with a wealth of impressive

exhibits and a year-round programme of visitor events. The Great Western Society, Didcot Railway Centre, Didcot, Oxon, OX11 7NJ. Tel. 01235 817200. www.didcotrailwaycentre.org.uk

3. AUNT SALLY'S SECRETS REVEALED

Vale and Downland Museum and Visitor Centre, Wantage

This friendly and innovative little museum provides a great introduction to the beautiful Vale of the White Horse, its small towns and enchanting villages, literary heritage and prehistoric sites. The Vale and Downland Museum and Visitor Centre, 19 Church Street, Wantage OX12 8BL. Tel. 01235 760176.

Tom Brown's School Museum, Uffington

Housed in a schoolroom, built in 1617, this volunteer-run museum features a wealth of fascinating material about local life, in particular the Uffington-born Thomas Hughes, author of *Tom Brown's Schooldays* and *The Scouring of the White Horse*. Tom Brown's School Museum, Broad Street, Uffington, Oxon SN7 7RA. www.museum.uffington.net. The annual Uffington White Horse Show, held over August Bank Holiday weekend, is quite an experience.

Coate Water Country Park

Noted for its diverse population of water birds, dragonflies and damselflies, the park is located on the south-eastern outskirts of Swindon, close to junction 15 of the M4. Great for walking and wildlife spotting. See it while the views towards the Wiltshire Downs are still gloriously uninterrupted by planned development. Swindon Services Ranger Team, Coate Water Country Park, Marlborough Road, Swindon SN3 6AA. Ranger centre tel. 01793 490150. www.swindon.gov.uk/natureforall

4. STEAMOPOLIS

STEAM – Museum of the Great Western Railway, Swindon, Wiltshire

A bold and imaginative display of railway relics and hands-on exhibits, offering glimpses of life 'inside' – as the Swindon engineering works was known locally. Don't miss the Wall of Names, Mr Brunel's stone collection and the silver coffee pot. Best visited on one of the regular Meet the Railway Workers days. Kemble Drive, Swindon, SN2 2TA. Tel 01793 466646. www.steam-museum.org.uk. Swindon's railway village is about ten minutes walk from here, via the pedestrian tunnel under the mainline, and well worth exploring.

National Trust Central Office, Swindon
The trust's new eco-friendly central office, a stone's throw from the STEAM museum, opened in July 2005, brings together under one roof 400 staff previously scattered across several sites between London and Cirencester. It is an impressive light and airy building, with a shop and café open to the public, and plenty of seating inside and out. The building is named Heelis – the married name of Beatrix Potter, one of the trust's patrons. National Trust Central Office, Heelis, Kemble Drive, Swindon SN2 2NA. Tel. 01793 817400. www.nationaltrust.org.uk

Science Museum, Wroughton, Wiltshire
Houses the Science Museum's large objects collection, everything from hovercrafts and early computers to MRI scanners and a Lockheed Constellation airliner, on a 545-acre site below the Wiltshire Downs. The Science Museum, Hackpen Lane, Wroughton, Swindon SN4 9NS. Tel. 01793 846200. www.sciencemuseum.org.uk/wroughton

5. BOX HILL & THE CORSHAM QUARRIES

Bowood House and Gardens, Wiltshire
A stately home set in parkland designed by Capability Brown. For the under twelves, there's an adventure playground, and a soft-play palace for the younger ones. Grown-ups can view Napoleon's death mask, Byron's Albanian costume, the Bowood Laboratory, several Old Masters and some fine watercolours, among other things. Further information from The Estate Office, Bowood, Calne, Wiltshire SN11 OLZ. Tel. 01249 812102. www.bowood.org

Corsham Tourist Information and Heritage Centre, Wiltshire
The subterranean secrets of Box Hill and the Corsham stone quarries during peacetime and the Second World War are revealed here, as well as some myths and legends, such as captured UFOs, and reserves of steam locomotives are said to have been stored here. Very Twilight Zone, very Wiltshire. Bring your own flask of tea – there are some benches near the town hall. Arnold House, High Street, Corsham, Wiltshire SN13 OEZ. Tel. 01249 714 660.

The William Herschel Museum, Bath
Celebrates the achievements of the star-struck William and Caroline Herschel, the founders of modern astronomy. They lived in the townhouse that now

accommodates the museum, before moving to Slough under the patronage of their greatest supporter, King George III. Typical perhaps of the women of the time, Caroline was unduly modest, comparing the supporting role she played in her brother's work to that of a puppy dog. In fact, she was an expert comet-hunter and recorder of activity in the night skies, as well as William's devoted housekeeper. Herschel House, 19 New King Street, Bath BA1 2BL. Tel. 01225 311342. www.bath-preservation-trust.org.uk/museums/herschel

Avon Valley Railway, Bristol
Train rides, special events and, from its new Avon Riverside station, the chance to explore the Avon River Valley – east of Bristol – by foot, bike or riverboat. Avon Valley Railway, Bitton Railway Station, Bath Road, Bitton, Bristol BS30 6HD. Tel. 0117 932 5538. www.avonvalleyrailway.org

6. CAPITAL WORKS

GWR FM Big Balloon, Bristol
A ride on this tethered helium balloon will take you nearly 500ft above Castle Park, Bristol. On clear days, the views stretch for 25 miles in each direction. www.gwrfmbigballoon.co.uk

SS Great Britain, Bristol
Launched in 1843 to provide luxury travel to New York, Brunel's famous ship was the world's first great ocean liner. Now fully restored at her original dock, open for self-guided tours and numerous special events. It's a marvel. Great Western Dockyard, Bristol BS1 6TY. Information line 0117 929 1843. www.ssgreatbritain.org

Clifton Suspension Bridge, Bristol
Brunel's first major commission, it was an extraordinary feat of engineering. Although designed in the nineteenth century for light horse-drawn traffic, the toll bridge still meets the demands of modern commuter traffic, with up to 12,000 vehicles crossing it daily. Those with a good head for heights can walk it; those without (like me) can view it from the Avon Walkway and cyclepath down on the western banks of the river. For information about visitor services, including guided tours at weekends, telephone 0117 974 4644 or email visitinfo@clifton-suspension-bridge.org.uk. For operational and administrative inquiries: Clifton Suspension Bridge Trust, Bridgemaster's Office, Leigh Woods, Bristol BS8 3PA.

Tyntesfield, Wraxall, Somerset
This spectacular Victorian house, home to four generations of the Gibbs family, was acquired by the National Trust in 2002 after a major fund-raising campaign. The rebuilding of Tyntesfield was undertaken by William Gibbs and completed in 1865, when he was well into his seventies. His brother, George Henry Gibbs, was a GWR director and stalwart ally of Brunel. Tyntesfield, Wraxall, Somerset BS48 1NT. Information line 0870 458 4500. www.nationaltrust.org.uk/tyntesfield

Clevedon Court
This medieval manor house was built by the Norman de Clevedon family, and was home to ten generations of the Elton dynasty. Wonderful eighteenth and nineteenth-century prints of railways, bridges and canals adorn the Queen Anne staircase. Don't miss Dame Dorothy Elton's collection of coloured-glass rolling pins and cucumber straighteners! Clevedon Court, Tickenham Road, Clevedon, Somerset BS21 6QU. From the National Trust home page www.nationaltrust.org.uk click on visits and holidays for the relevant link to more information.

Clevedon Pier
The most beautiful pier in England, according to Sir John Betjeman, lovingly restored and maintained by enthusiasts. For details of opening times, events and facilities, contact the Clevedon Pier and Heritage Trust, The Toll House, Clevedon Pier, North Somerset BS21 7QU. Tel. 01275 878846. www.clevedonpier.com. Summer cruises can be made from the pier. Timetables and other cruise details are available from Waverley Excursions Ltd, Waverley Terminal, 36 Lancefield Quay, Glasgow G3 8HA. Tel. 0845 130 4647 www.waverleyexcursions.co.uk

7. THE FASTEST WAY TO SLOW DOWN

West Somerset Railway
A preserved railway running steam trains from Bishops Lydeard, near Taunton, through delightful countryside to the coast at Minehead 20 miles north. Travel in a 1950s coach painted in the distinctive chocolate and cream livery associated with the GWR and alight at any one of ten stations beautifully restored and maintained by volunteers, where there's plenty to do and see. West Somerset Railway, Minehead, Somerset TA24 5BG. Tel. 01643 704996. www.West-Somerset-Railway.co.uk

Grand Western Horseboat Company, Tiverton, Devon

The Grand Western Canal was built in 1814, primarily for the lime trade, and worked by horse-drawn tub-boats. Today the canal is a country park owned and run by Devon County Council with free access all year round. Like the Bridgwater and Taunton Canal, it's a delightful oasis of calm, abundant in flora and fauna with sumptuous country views. You can walk or cycle along the canal, or from March to October, splash out on a trip in a horse-drawn barge, one of only a handful left in England. From the canal shop at Tiverton Wharf, you can get refreshments and hire boats. Grand Western Horseboat Company, The Wharf, Canal Hill, Tiverton, Devon. Tel. 01884 253345. www.horseboat.co.uk

Devon Railway Centre, Bickleigh, Devon

Abandoned to the elements for thirty-five years, the old Cadleigh and Bickleigh station has been restored and given a new lease of life by a group of dedicated enthusiasts as a centre for both rail buffs and families seeking an entertaining day out. Take train rides on a narrow gauge and a miniature railway, marvel at the model railway layouts and a museum collection and visit the café in the old waiting room. Not to be confused with the South Devon Railway, of which more later. Devon Railway Centre, Bickleigh, Tiverton, Devon EX16 8RG. Tel. 01884 855671. www.devonrailwaycentre.co.uk

8. ATMOSPHERIC PRESSURE ON THE ENGLISH RIVIERA

Dawlish Warren National Nature Reserve

A blissful spit of land which juts out into the mouth of the River Exe, an area of sand dunes, mudflats and beach, rich in wildlife – an important stop-over and feeding ground for many thousands of migrating wildfowl and wading birds from August to late March. It's a great for walking, but watch out for flying golf balls near the bird hide. There's a visitor centre and car park. The Wardens, Dawlish Warren NNR, Teignbridge District Council, Forde House, Brunel Road, Newton Abbot TQ12 4XX. Tel. 01626 863980 or 01626 215754. www.teignbridge.gov.uk/countryside

Newton Abbot Town and GWR Museum

A charming little museum, tucked away in a quiet street. A lot is packed into the GWR room, including Mr Brunel's ornately fashioned toilet, and opportunities to work railway-signalling equipment. The walls are festooned with wonderful old illustrations, notably a huge poster of the Royal Albert

Bridge, Saltash. Good displays on the doomed atmospheric system, and the rise of 'Little Swindon'. Small but perfectly formed. Newton Abbot Town & GWR Museum, 2a St Paul's Road, Newton Abbot, Devon TQ12 2HP. Tel. 01626 201121.

Brunel Manor
Stands on the site where Brunel planned to build his retirement home. Only the foundations were laid before his death. Now a hotel and conference centre owned by the Woodland House of Prayer Trust, the manor welcomes Christian and non-Christian groups. Brunel Manor, Teignmouth Road, Torquay, Devon RQ1 4SE. Tel. 01803 329333. www.brunelmanor.com

Cliff Railway, Babbacombe
More than 250,000 passengers are carried each season on this 720ft long railway connecting Oddicombe Beach and Babbacombe. It's been operating since 1926 (apart from during the Second World War), and runs daily from Easter to September. Torbay Council has been seeking a private operator to take over the running of the railway. Babbacombe Cliff Railway, Environment Services, Torbay Council, Civic Offices, Castle Circus, Torquay TQ1 3DR. Tel. 01803 201201.

Paignton and Dartmouth Steam Railway
Steam train rides through fabulous scenery between Paignton and Kingswear, with the option of going on from Kingswear to Dartmouth on the company's ferry. Other combined ticket options include an additional cruise along the Dart; and a round-robin trip of train, ferry ride, boat cruise to Totnes and then a bus back to Paignton or Goodrington. A great little integrated transport system operated by Dart Valley Railway PLC. Tel. 01803 555872. www.paignton-steamrailway.co.uk

South Devon Railway
This railway hugs the attractive River Dart all the way from Totnes Littlehempston station (a short, pleasant stroll from Totnes main-line station) to Buckfastleigh, 7 miles up river, along a former GWR branch line. Places to visit before or after a train ride include, at Buckfastleigh, the railway museum, Buckfast Abbey (via vintage bus), a butterfly farm and otter sanctuary; and at Totnes Littlehempston a rare breeds centre and the historic town of Totnes itself. South Devon Railway, The Railway Station, Buckfastleigh, Devon TQ11 ODZ. Tel. 0845 345 1420. www.southdevonrailway.org

9. INTO A CORNISH AUTUMN

Royal Albert Bridge, Saltash, Cornwall

The Royal Albert (rail) and the Tamar (road) bridges span the river between St Budeaux, on the north-western outskirts of Plymouth, Devon and Saltash, Cornwall. There are many good spots for viewing the bridges: a car park and interpretation board near the tollbooths on the St Budeaux side; and at Saltash, the railway station platform and the Waterside area provide good vantage points. The Saltash Heritage Trail is signed at several points around the eastern side of the town. More information from www.saltash.gov.uk – follow the links from the home page to 'partnership' and 'projects'. The Tamar Valley Line from Plymouth main-line station to Gunnislake via the famous Calstock viaduct is said to be among the most scenic railway routes in England.

Railholiday Ltd, Cornwall

This company offer self-catering holiday accommodation in converted railway vehicles. A passenger luggage van, built for the London and South Western Railway and which entered service in 1896, is sited in a siding near the main line and the owners' home at St Germans. A 1957 British Rail corridor carriage is situated in a former goods yard at Hayle, near St Ives Bay. A third rail vehicle, a refurbished travelling post office, is due to be offered for letting from 2006. Railholiday Ltd, Haparanda Station, Nut Tree Hill, St Germans, Cornwall PL12 5LU. Tel. 01503 230 783. www.railholiday.co.uk

Bodmin and Wenford Railway

Steam train rides between Bodmin Parkway station and Boscarne, linked with walking and cycling routes, notably the celebrated Camel Trail. The lovingly restored GWR station at Bodmin General is the line's HQ, and you can hire bikes both here and at Wadebridge, further along the Camel Trail towards Padstow. Lanhydrock House (National Trust) lies a pleasant 45-minute largely wooded walk from Bodmin Parkway station. Bikes and dogs conveyed on trains free of charge. Bodmin and Wenford Railway, Bodmin General Station, Bodmin, Cornwall PL31 1AQ. Tel. 0845 1259678. www.bodminandwenfordrailway.co.uk

John Betjeman Centre, Wadebridge, Cornwall

In search of a sandwich and somewhere to water and rest the dog, I stumbled across the old Wadebridge station, now housing an excellent day centre for the recreation and education of senior citizens; and a John Betjeman

Memorabilia Room, displaying personal mementos, academic honours and furniture belonging to the late Poet Laureate, who lived nearby and is buried at St Enodoc Church. Well-behaved dogs, and humans, welcome. The John Betjeman Centre, Southern Way, Wadebridge, Cornwall PL27 7BX. Tel. 01208 812392. The building is owned by Wadebridge Concern for the Aged, and run by volunteers.

10. EXPLORING THE COPPER KINGDOM

China Clay Country Park, Mining and Heritage Centre

Here you can see a working china clay pit, and discover the fascinating history of what remains Cornwall's largest industry. Plenty to do and see for all ages under cover and outdoors. China Clay Country Park, Wheal Martyn, Carthew, St Austell PL26 8XG. Tel. 01726 850362. www.chinaclaycountry.co.uk

National Coastwatch Institution

Two fishermen lost their lives off the Cornish coast within sight of a recently closed coastguard station – Government cuts had shut many small coastguard stations around the country. A voluntary organisation, the NCI, was set up in 1994 in the wake of the Cornish tragedy in order to restore a visual daylight watch to UK shores. The NCI's volunteer force has since grown to more than 1,300 people operating 28 lookouts on a rota basis, and working closely with the Maritime Coastguard Agency, emergency services and other public bodies. NCI volunteers have helped to ensure dozens of successful rescue missions around our coastline. More lookouts are planned subject to resources and availability of suitable sites. www.nci.org.uk

Bibliography & Further Reading

Christie, Agatha, *An Autobiography*, Collins, 1977

Clevedon Court, National Trust, 2003

Coombes, Nigel, *Striding Boldly: The Story of Clevedon Pier*, Clevedon Pier Trust, 1995

Delicato, Aldo and Cole, Beverley, *Speed to the West: Great Western Publicity and Posters, 1923–1947*, Capital Transport, 2000

Harper, Charles George, *From Paddington to Penzance: The Record of a Summer's Tramp from London to the Land's End*, Chatto & Windus, 1893

Oakley, Mike, *Wiltshire Railway Stations*, Dovecote Press, 2004

Phillips, Daphne, *The Great Road to Bath*, Countryside, 1983

Simmons, Jack (ed.), *The Birth of the Great Western Railway: Extracts from the Diary and Correspondence of George Henry Gibbs*, Adams & Dart, 1971

Tyntesfield, National Trust, 2003

Vaughan, Adrian, *Isambard Kingdom Brunel*, John Murray, 1991

Williams, Alfred, *Life in a Railway Village*, revd edn, Sutton, 1984

ORDNANCE SURVEY EXPLORER MAPS

173 London North

172 Chiltern Hills East

160 Windsor Weybridge & Bracknell

171 Chiltern Hills West

159 Reading, Wokingham & Pangbourne

170 Abingdon, Wantage & Vale of White Horse

169 Cirencester & Swindon

156 Chippenham & Bradford-on-Avon

155 Bristol & Bath

154 Bristol West

153 Weston-super-Mare

140 Quantock Hills & Bridgwater

128 Taunton & Blackdown Hills

114 Exeter & the Exe Valley

110 Torquay & Dawlish

OL28 Dartmoor

OL20 South Devon

108 Lower Tamar Valley & Plymouth

107 St Austell & Liskeard

106 Newquay & Padstow

105 Falmouth & Mevagissey

104 Redruth & St Agnes

103 The Lizard

102 Land's End

ORDNANCE SURVEY LANDRANGER MAP

175 Reading & Windsor